THE
Stannic Oxide Gas Sensor

PRINCIPLES and APPLICATIONS

T0179091

THE
Stannic Oxide
Gas Sensor
PRINCIPLES and APPLICATIONS

Kousuke Ihokura

Joseph Watson

CRC Press
Taylor & Francis Group
Boca Raton London New York

CRC Press is an imprint of the
Taylor & Francis Group, an **informa** business

CRC Press
Taylor & Francis Group
6000 Broken Sound Parkway NW, Suite 300
Boca Raton, FL 33487-2742

© 1994 by Taylor & Francis Group, LLC
CRC Press is an imprint of Taylor & Francis Group, an Informa business

First issued in paperback 2019

No claim to original U.S. Government works

ISBN 13: 978-0-367-44951-3 (pbk)
ISBN 13: 978-0-8493-2604-2 (hbk)

**Visit the Taylor & Francis Web site at
http://www.taylorandfrancis.com**

**and the CRC Press Web site at
http://www.crcpress.com**

Library of Congress Cataloging-in-Publication Data

Ihokura, Kousuke.
    The stannic oxide gas sensor : principles and applications / authors,
    Kousuke Ihokura, Joseph Watson.
        p.   cm.
    Includes bibliographical references and index.
    ISBN 0-8493-2604-4
    1. Gas-detectors.   2. Stannic oxide.   I. Watson, J. (Joseph)
    II. Title.
TP754.I36 1994
681'.2—dc20

                                                              93-35554
                                                              CIP

Library of Congress Card Number 93-35554

PREFACE

Over the past few decades, the solid state gas sensor based on stannic oxide (also known as tin dioxide or tin IV oxide) has become the predominant solid-state device for gas alarms used on domestic, commercial, and industrial premises. This is because it is perceived as a long-lived, low-cost sensor which requires only minimal and, therefore, reliable electronics, so that little or no maintenance is involved.

However, the principles behind its operation are by no means simple and a basic understanding of them can lead to much enhanced methods of utilization. The present book has, therefore, been designed to offer a comprehensive account of these principles, but in largely qualitative terms, along with details of various methods of fabrication, and both have been related to observed characteristics.

Methods of modifying these characteristics are also presented, as are the basic design tenets of the necessary associated circuits. Hence, the book should be of interest to scientists who wish a fundamental grounding in the topic, to technologists involved in the fabrication of the relevant sensors, and also to the engineers responsible for their application. At all points, sufficient references have been included so that the reader may pursue any particular aspect in greater detail.

We also demonstrate that improvements in the technology and fabrication of the stannic oxide based sensor are leading to improved stability and selectivity, which are necessary for its use in gas concentration monitors in addition to the earlier simple alarm instruments.

Finally, some methods of testing and characterization are presented, because national and international legislation will progressively demand ever-increasing sophistication in such techniques and their utilization.

K.I.
J.W.
September 1993

TABLE OF CONTENTS

Introduction

1.   THE METAL OXIDE GAS SENSOR

Of several possible metal oxide gas sensors, only those based on stannic oxide in its ceramic form have become widely manufactured and utilized. It is therefore the purpose of this book to describe the chemical and physical processes underlying the operation of this form of sensor, various methods of manufacture, the resulting electrical and environmental characteristics and finally the nature of the relevant electronic circuits which make operation possible. Hence, the book is oriented not only towards the investigative science community, but also to engineers, who are responsible for the manufacture and practical application of such sensors.

2.   THE NATURE OF SENSORS

A 'transducer' is a device which converts one form of energy into another, and a 'sensor' is a form of transducer which converts a physical or chemical quantity into an electrical quantity for purposes of measurement. In particular, a chemical sensor — which includes the stannic oxide gas sensor — is intended to determine the composition and concentration of the relevant material via an electrical signal. It will be shown that the stannic oxide sensor can to some extent differentiate between various gases, but is much more successful at determining the concentrations of known gases, and does so by undergoing a resistance change which can be easily detected and either monitored or used to actuate an alarm or controller.[1]

Terminologically, the active part of a complete gas sensor assembly is called the sensor element, whilst a complete instrument incorporating such a sensor can take the form of a gas monitor, gas detector or gas alarm.

3.   EARLY GAS SENSORS

The earliest commercial gas sensor appeared about 1923 and consisted of a hot platinum wire: indeed, the hot wire sensor is still in use today.[2] If such a filament is maintained at several hundred degrees centigrade above ambient, then via catalysis it will detect any combustible gas which may be present in the atmosphere, resulting in a rise in temperature in the platinum wire and a concomitant resistance increase which may be measured. This is usually done via a resistance bridge circuit which typically provides an output signal of only a few millivolts resulting from (for example) about 1000 ppm of isobutane.

In 1959, this sensor was improved by Baker,[3] whose 'catalytic combustion sensor' involved a catalyst such as palladium carried by an α-alumina substrate

1

in the shape of a bead. Though the accompanying circuit did produce some 15–20 mV, it was not easy to overcome the instability of the sensor with time which resulted from deterioration of the catalyst. Usage in coal mines, for which the sensor had been designed, was therefore limited, and thermistor methods were introduced about 1961. However, modern versions of the catalytic sensor — often known as Pellistors — are now comparatively robust and easily obtainable.

Again in the 1960s, gas explosions became frequent in Japan, and many were related to the popularization of bottled LP gas as a domestic source of energy for heating and cooking. Hence, a requirement arose for a sensitive and inexpensive sensor which could reasonably be used in the domestic environment.

In 1962, Seiyama et al.[4] reported studies of thin film zinc oxide gas sensors based on the semiconductor catalysis mechanism, and this was the first device which utilized the resistance change in a metallic oxide semiconductor which results from gas adsorption.[5]

This development was the precursor to major efforts in the field by various groups around the world. In particular, Taguchi patented a stannic oxide ceramic gas sensor[6] in October of 1962 which led directly to the eventual mass production of the world's first commercial devices (in 1968) and this was subsequently augmented by the discovery of the sensitivity-improving properties of noble metal inclusions and the progressive optimization of sintering procedures.

World-wide work on metal oxide sensors has continued and has been extended to other materials; and the range of detectable gases has expanded from combustible species to oxygen, the oxides of nitrogen, hydrogen sulphide and many others. Table 1 lists a series of key papers in various important aspects of research which took place in the two decades following Seiyama's initial work on zinc oxide in 1962.

4.   A HISTORICAL NOTE

It is in Japan where the major production and commercialization of gas sensors has taken place, and there the market grew to some 150,000 units per month by 1977. This led to further research, which in turn led to the organization of the Sensor Research Group of the Electrochemical Society of Japan, again in 1977. This has now expanded to encompass sensors for chemical species other than gases, and the generic term, 'chemical sensor' is now in common usage.

The concept of the chemical sensor was widely disseminated at the first International Meeting on Chemical Sensors in Fukuoka and at the Second International Conference on Solid-state Sensors and Actuators in the Netherlands, both held in 1983. At this latter conference, the chemical sensor appeared to have eclipsed the physical sensor for the first time in the volume of reported work.

5.   CURRENT RESEARCH AND DEVELOPMENT

During the decade of the 1970s, research on solid-state gas sensors concentrated firstly on a search for the best material, and secondly upon the nature of additives for sensitivity enhancement. In the 1980s, this groundwork was used

TABLE 1. Key Solid-State Gas Sensor R & D Papers in the First Two Decades

Year	Material	Signal	Detecting gas	Researcher	Ref.
1962	ZnO (Thin film)	E.C.	H^2, Alcohol, etc.	Seiyama, et. al.	4
	SnO_2	E.C.	Combustible gas	Taguchi.	6
1963	SnO_2+Pd, Pt, Ag.	E.C.	Combustible gas	Taguchi.	7
1966	ZnO, SnO_2, etc.	E.C.	Reducing gas	Seiyama, et. al.	8
	SnO_2+Al_2O_3	E.C.	Combustible gas	Taguchi.	9
1967	WO_3+Pt	E.C.	H_2, N_2H_4, WH_3, H_2S	Shaver.	10
	In_2O_3+Pt	E.C.	H_2, Hydrocarbon	Loh.	11
1969	SnO_2+SiO_2	E.C.	Combustible gas	Taguchi.	12
1971	ZnO+Pt+Ga_2O_3	E.C.	CH_4, NH_3	Bott, et. al.	13
1972	SnO_2+Pd	E.C.	Propane	Seiyama, et. al.	14
1975	$La_{1-x}Sr_xCoO_3$, etc.	E.C.	Alcohol	Sakurai, et. al.	15
	V_2O_5+Ag (Thin film)	E.C.	NO_2	Sakai, et. al	16
	ZnO+Ga_2O_3+Pd, Pt.	E.C.	H_2, CO, Hydrocarbon	Ichinose, et. al.	17
	TiO_2	E.C.	O_2	Tien, et. al.	18
	CoO	E.C.	O_2	Logothetis, et. al.	19
	Pd MOS FET	T.V.	H_2	Lundström.	20
1976	Pd/CdS	R.	H_2	Steele, et. al.	21
1977	SnO_2+Pd+ThO_2	E.C.	CO	Nitta, et. al.	22
1978	γ-Fe_2O_3	E.C.	Propane	Matsuoka, et. al.	23
	Co_3O_4	E.C.	CO	Stetter.	24
	Ag_2O	S.P.	Mercaptan	Tsubomura, et. al.	25
	Pd/TiO_2	R.	H_2	Tsubomura, et. al.	26
	Metal-phthalocyanine	E.C.	NO_2	Sadaoka, et. al.	27
	Anthracene	E.C.	Amine, Carboxylic acid	Suzuki, et. al.	28
1979	ZnO (Thin film)	E.C.	Alcohol	Heiland, et. al.	29
1980	SnO_2-Ultra fine particle	E.C.	Combustible gas	Abe, et. al.	30
1981	α-Fe_2O_3	E.C.	CH_4, H_2, etc.	Nakatani, et. al.	31
	ZnO+V_2O_5+MoO_3	E.C.	Freon	Shiratori, et al.	32
1982	ZnO (Single crystal)	E.C.	CO	Jones, et. al.	33
	SnO_2 (Thin film)	E.C.	Combustible gas	Chang.	34
	SnO_2 (Thin film)	E.C.	Combustible gas	Sotomura, et al.	35

E.C.: Electric conductivity. T.V.: Threshhold voltage. R.: Rectification. S.P.: Surface potential.

to develop a series of commercial sensors to cover a very wide variety of applications.

More recently, several original studies have commenced, including some involving the use of a number of sensors having different characteristics, the outputs of which are combined by sophisticated signal-conditioning circuitry to result in a gas identification system.[36] This technique overcomes the rather limited selectivity performance of any individual sensor, but has the disadvantage that if any sensor is changed, some re-programming of the signal-conditioning section may well become necessary because even modern sensors do have rather wide parameter tolerances.

Improvements in the selectivity of individual sensors are taking place, however, and techniques other than the employment of various additives are under investigation. For example, selectivity can be improved by cycling the sensor temperature between limits,[37] and this method has resulted from the development of carbon monoxide sensors for the detection of fires in their very early stages prior to the operation of smoke detectors. It has also been recently shown that non-specific sensors can be used for fire detection via the production of hydrogen,[38] which often occurs along with carbon monoxide and which may reach a sensor first because of its greater diffusion rate.

Another approach to selectivity involves surface modification of the base metal oxide with hydrophobic groups,[39] calcium oxide,[40] super-corpuscles of gold,[41] zinc oxide,[42] cuprous oxide,[43] sulphur,[44] or lanthanum oxide.[45] This approach differs from those involving additives such as antimony or vanadium which operate via valence control, or the noble metal additives which involve catalysis.

Earlier sensors were useful for the detection of general combustible gases such as methane or carbon monoxide, but modern work is tending towards the development of sensors for specific purposes such as the detection of low concentrations of gases such as hydrogen sulphide,[39,42,43] the nitrogen oxides[47] and mercaptans[42] in addition to general air pollutants.[46] Somewhat exotic application-driven developments include sensors for trimethylamine to determine the freshness of fish,[48] for general cooking gases in the control of microwave ovens[49] and for the detection of leaking Freon and other refrigerants.[44] It will be recognised that the detection of some of these gases is relevant to the current environmental concerns involving ozone destruction in the upper atmosphere.

In the future, very modern developments such as material design at the molecular level and micromachining technology will undoubtedly be applied to solid-state sensors, as will contemporary electronic drive and monitoring methods, and the concatenation of sensors with application specific integrated circuits (ASICs).

However, at the time of writing, only discrete gas sensors based on stannic oxide have resulted in a major market (currently at some 600,000 units per month), and R & D work continues on this material, which has clearly yet to realize its full potential. The present book is therefore devoted to this material, for as has been mentioned, it is intended not only as a research review, but as an engineering introduction to this form of sensor and its applications.

6.   THE TAGUCHI GAS SENSOR (TGS)

Because by far the greatest production and utilization of stannic oxide gas sensors is centered in Japan, where Taguchi was responsible for the first patents in 1962, it is appropriate to include a short history of the subject in that country. Furthermore, most of the sensors and sensor materials used as examples in the text stem from the relevant firm, Figaro Engineering Inc., and the type numbers (such as the TGS 203 etc.) reflect their historical origin.

It is in Japan where bottled (LP) gas is most popular, being used by some twenty-three million households for their cooking requirements; and piped gas is used by at least another eighteen million. Thus, of all developed countries, Japan is the prime user of gas, but concomitantly suffers the most gas explosion accidents.

The greatest proliferation in bottled gas supply occurred around 1962, and as the accident rate built up, this became a social problem of much greater magnitude than seen in other countries such as the United Kingdom, where the Ronan Point disaster highlighted a similar problem, albeit involving only piped gas. It was this situation which led Taguchi to develop and patent the stannic oxide sensor, and he subsequently established the Figaro Engineering Co. Inc. to produce the sensors commercially. Production actually began in 1968 after gas sensitivity improvements by noble metal additives had been demonstrated in 1963, followed by the adoption of alumina sintering aggregate which resulted in great ceramic strength and a further increase in sensitivity. Hence, a low-cost but sensitive and robust domestic gas sensor was born.

Shortly after mass production was under way, ethyl silicate was introduced as a binder, which yet again improved the mechanical strength of the active material, and hence its reliability and longevity. These and other techniques will be detailed in the following text.

The early domestic gas sensors led to a striking reduction in bottled gas explosion accidents in Japan, as shown in Figure 1,[50] though other countries

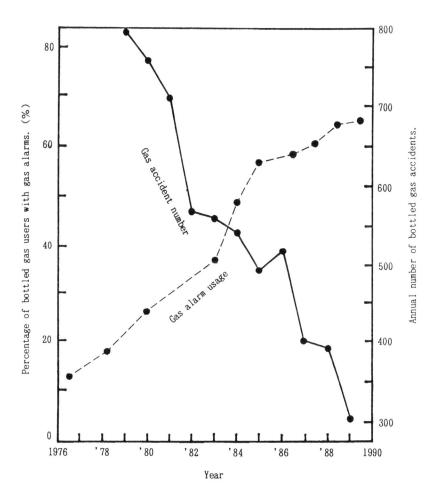

FIGURE 1. The fall in bottled gas accidents compared with the rise in gas alarm usage in Japan.[50] ●———●: Annual number of bottled gas accidents in Japan. ●----●: Percentage of bottled gas users with gas alarms. (From Gas Alarm Industries Assoc. Jpn., May 1950.)

were slow to take up such devices. There were various reasons for this, one being that the early Japanese circuits were designed for the 100 V Japanese mains supply and did not transfer conveniently to the 220–240 V European system. Also, the national gas bodies in those countries could not easily relate such cheap and simple domestic gas detectors to the terms of their own regulations and guidelines which were, and are, stringent. This latter caveat also applied in the U.S.A. even though the mains voltage there is only 110 V. However, locally-manufactured detectors for recreational vehicles (RVs) and boats did establish a small but continuing market.

In recent years, industrial versions of stannic oxide gas sensors have been developed and these have become widely accepted world-wide, in part because their very robust encapsulations and closely-controlled active element compositions have made possible their use in a very wide range of industrial environments including the harshest.

This and many other developments have taken place within the laboratories of the Figaro company,[51] whose products now comprise some 95% of the world market. The Japanese author was associated with Taguchi for some years before joining that company in 1973, which is why most of the sensor examples given in the text have the characteristic 'TGS' numbers; and the British author has been associated with gas sensor applications and electronics in both Europe and the U.S.A. over the same period and has co-operated with the Figaro company for much of that time.[52] It is therefore hoped that the book will serve as at least an introduction to this burgeoning topic for both scientific (research) and engineering (applications) readerships, and that it will lead to a greater understanding of the unique advantages and limitations of this form of gas sensor.

REFERENCES

1. Seiyama, T., et al, "Kagaku Sensa" (Chemical Sensors), chap.1, (Kodansha, 1982).
2. Egashira, M., Sessyoku-nensyo-shiki Sensa (Catalytic Combustion Gas Sensors), in *Kagaku Sensa Jitsuyo Benran (Practical Handbook of Chemical Sensors)*, Sec.7, Fuji Tekunoshisutemu, 1986, 61.
3. Baker, A.R., Brit. Pat. 892, 530.
4. Seiyama, T., et al, A new detector for gaseous components using semi-conductive thin films, *Anal. Chem.*, 34, 1502, 1962.
5. Tarama, K., On the catalysis of the transition metal oxide, Yukigosi Kagaku (J. Synth. Org. Chem. Japan), 16, 433, 1958.
6. Taguchi, N., Japanese Pat. S45-38200 (appl. for 1962).
7. Taguchi, N., Japanese Pat. S47-38840 (appl. for 1966).
8. Seiyama, T., et al, Study on a detector for gaseous components using a semiconductive thin film, *Anal. Chem.*, 38, 1069, 1966.
9. Taguchi, N., Japanese Pat. S50-30480 (appl. for 1966).
10. Shaver, P.J., Activated tungsten gas detectors, *Appl. Phys. Lett.*, 11, 255, 1967.

11. Loh, J.C., French Pat. 1545292 (appl. for 1967).
12. Taguchi, N., Japanese Pat. S50-23317 (appl. for 1969).
13. Bott B., et al, British Pat. 1 374 575 (appl. for 1971).
14. Seiyama, T., Gas detection by activated semiconductor gas sensor, *Denki Kagaku (J. Electrochem. Soc. Jpn.)*, 40, 244, 1972.
15. Sakurai, H., et al, A gas sensor made of rare earth transition metal oxide in perovskite structure, Material for Study Group on Elec. Devices, Inst. Elec. Eng. Japan, EDD-75-48, 1975.
16. Nakagawa, M., et al, Adsorption field-effect transistor, Material for Study Group on Elec. Devices, Inst. Elec. Eng. Japan, EDD-75-50, 1975.
17. Ichinose, N., et al, High performance gas sensor, Material for Study Group on Elec. Devices, Inst. Elec. Eng. Japan, EDD-75-53, 1975.
18. Tien, T.Y., et al, TiO_2 as an air-fuel ratio sensor for automobile exhausts, *Amer. Ceram. Soc. Bull.*, 54, 280, 1975.
19. Logothetis, E.M., et al, Oxygen sensors using CoO ceramics, *Appl. Phys. Letters*, 26, 209, 1975.
20. Lundstrom, I., et al, A hydrogen-sensitive MOS field-effect transistor, *Appl. Phys. Lett.*, 26, 55 (1975).
21. Steele, M.C., et al, Palladium/cadmium sulfide Schottky diodes for hydrogen detection, *Appl. Phys. Lett.*, 28, 687, 1976.
22. Nitta, M., et al, Oscillation phenomenon in ThO_2-doped SnO_2 exposed to CO gas, *Appl. Phys. Lett.*, 32, 590 (1978).
23. Matsuoka, M., et al, Gas sensitivity characteristics of γ-Fe_2O_3 ceramic, Material for Study Group on Elec. Devices, Inst. Elec. Eng. Japan, EDD-78-22, 1978.
24. Stetter, J.R., A surface chemical view of gas detection, *J. Colloid Interface Sci.*, 65, 432, 1978.
25. Yamamoto, N., Influence of chemisorption of 2-propanethiol on the electrical properties of silver oxide compaction, *Bull. Chem. Soc. Jpn.*, 54 , 696, 1981.
26. Sotomura, S., et al, Metal-semiconductor junctions for the detection of reducing gases and the mechanism of the electrical responses, *Nippon Kagaku Kaishi (J. Chem. Soc. Jpn.)*, 1980, 1585, 1980.
27. Sadaoka, Y., et al, The NO_2 detecting ability of the phthalocyanine gas sensor, *Denki Kagaku (J. Electrochem. Soc. Jpn.)*, 48, 486, 1980.
28. Suzuki, K., et al, Detection of gases using an organic semiconductor, anthracine, *Bunseki Kagaku (J. Jpn. Soc. Anal. Chem.)*, 27, 472, 1978.
29. Schultz, V.M., et al, Messung von fremdgasen in der luft mit halbleitersensoren (Measurement of extraneous gases in air by means of semiconducting sensors), *Technisches Messen tm (Germany)*, 1979, no.11, 405, 1979.
30. Abe, J., et al, Gas-sensitive sensor with ultrafine particle film deposited on monolithic IC chip, National Technical Report *(Tech. J. Matsushita Elec. Ind. Co. Ltd.)*, 26, 457, 1980.
31. Nakatani, K., et al, α-Fe_2O_3 ceramic gas sensor, Tech. Digest, 1st. Sensor Symposium, Inst. Elec. Eng. Japan, Tsukuba, 1981, 25.
32. Shiratori, M., et al, Gas sensors for halogenated hydrocarbons, Tech. Digest, 1st Sensor Symposium, Inst. Elec. Eng. Japan, Tsukuba, 1981, 25.

33. Jones, T.A., et al, Solid-state sensors: zinc oxide and phthalocyanine, Abs. Int. Seminar on Solid-state Gas Sensors (23rd WEH Seminar), Bad Honnef, Germany, 1982, 7.
34. Shin-Chia Chang, Thin film tin oxide gas sensor, Abst. Int. Seminar on Solid-state Gas Sensors (23rd WEH Seminar), Bad Honnef, Germany, 1982, 7.
35. Sotomura, T., et al, Gas sensitivity of SnO_2 sputtered film, Digest of 2nd. Chemical Sensor Symposium, Electrochem. Soc. Japan: Assoc. Chem. Sens., Tokyo, 1982, 28.
36. Ikegama, A., et al, Olfactory detection using integrated sensor, Proc. Transducers '85 (3rd Int. Conf. on Solid-state Sensors & Actuators), Philadelphia, 1985, 136.
37. Murakami, N., The selective detection of carbon monoxide with stannic oxide gas sensor using sensor temperature cycle, Digest of 5th Chemical Sensor Symposium, Electrochem. Soc. Japan : Japan. Assoc. Chem. Sens., Tokyo, 1982, 53.
38. Amamoto, T., et al, A fire detection experiment in a wooden house by SnO_2 semiconductor gas sensor, Proc. Transducers '89 (5th Int. Conf. on Solid-state Sensors & Actuators), Montreux, 1989, 226.
39. Kanefusa, S., et al, High sensitivity H_2S gas sensors, *J. Electrochem. Soc.,* 132, 1770, 1985.
40. Fukui, K., et al, Alcohol-selective gas sensor, Digest of 4th Chemical Sensor Symposium, Electrochem. Soc. Japan : Japan. Assoc. Chem. Sens., Tokyo, 1985, 51.
41. Haruta, M., et al, A novel CO sensing semiconductor with co-precipitated ultrafine particles of gold, Proc. 2nd Int. Meet. on Chem. Sens., Bordeaux, 1986, 179.
42. Nakahara, T., et al, High sensitive SnO_2 gas sensor I. Detection of volatile sulphides, Proc. Symp. Chem. Sens., 1987 Joint Congress, Electrochem. Socs. Japan & U.S.A., Honolulu, 1987, 55.
43. Maekawa, T., et al, H_2S detection using $CuO\text{-}SnO_2$ sensor, Chemical Sensors, *(J. Jpn. Assoc. Chem. Sens.),* 6, suppl. B, p.21, Tokyo, 1990.
44. Nomura, T., et al, Development of semiconductor Freon gas sensor, Chemical Sensors, *(J. Jpn. Assoc. Chem. Sens.),* 6, suppl. B, p.13, Tokyo, 1990.
45. Matsushima, S., et al, Role of additives on alcohol sensing by semiconductor gas sensor, *Chem. Letters (Chem. Soc. Jpn.),* 1989, 845, 1989.
46. Shiratori, M., et al, A thin film gas sensor for ventilation, Digest of 4th. Chemical Sensor Symposium, Electrochem. Soc. Japan : Japan. Assoc. Chem. Sens., 6, supp. B, Tokyo, 1985, 57.
47. Satake, K., et al, NO_x sensors for exhaust monitoring, Digest of 9th. Chemical Sensor Symposium, Electrochem. Soc. Japan : Japan. Assoc. Chem. Sens., Tokyo, 1989, 97.
48. Takao, Y., et al, The detection of trimethylamine by Ru-doped semiconductor gas sensor, Digest of 6th Chemical Sensor Symposium, Electrochem. Soc. Japan : Japan. Assoc. Chem. Sens., J. Electrochem. Soc. Japan, Tokyo, 1989, 23.

49. Ono, K., et al, Sensors for automatic cooking, Digest of 7th Chemical Sensor Symposium, Electrochem. Soc. Japan : Japan. Assoc. Chem. Sens., Urawa, 1988, 173.
50. Publicity material, Gas Alarm Industries Assoc. of Japan, May 1990.
51. Chiba, A., Development of the TGS gas sensor, *Chem. Sens. Tech.,* 4 , 1, 1992.
52. Watson, J., Ihokura K., and Coles, G.S.V., The tin dioxide gas sensor, *Inst. Phys. Meas. Sci. Technol.,* 4, 711, 1993.

CHAPTER ONE

Fabrication of the Stannic Oxide Ceramic Sensor

1.1 PREPARATION OF STANNIC OXIDE POWDER

The stannic oxide crystal has a rutile structure belonging to the tetragonal system whose lattice constants are a = 4.72 Å and c = 3.16 Å. The highest melting point reported is 2250°C and the crystal has no phase transition below this melting point.[1] Crystalline stannic oxide exhibits n-type semiconductor characteristics, and its donor level is formed by oxygen vacancy defects in the lattice.[2]

Ceramic gas sensor elements are fabricated by sintering stannic oxide powder, and the methods of preparing this powder markedly affect the degree of crystallinity, the crystallite size, the density of lattice defects and the crystal surface area and surface structure. These in turn determine the characteristics and performance of the eventual gas sensors, and so are commercially sensitive factors.

According to protocols established by Figaro Engineering Inc., stannic oxide may be prepared by dissolving tin of high purity in acid, followed by the addition of an alkaline solution to precipitate tin hydroxide, which is finally calcined to obtain the stannic oxide.

Another method of preparation, also described by Figaro Engineering Inc.,[3] involves the addition of 28% ammonia solution to 85% tin chloride solution. The tin hydroxide precipitate so obtained is separated by decanting and drying at 200°C for 4 hours and is then calcined in air in an electric furnace for 1 hour at 450°C to obtain stannic oxide.

Stannic oxide powders having different crystallite sizes and surface areas can be produced by changing the temperature and time of calcination. A noble metal inclusion may be added in the form of an aqueous solution and dried at 200°C for 1 hour before calcination, after which the material may be fully powdered in a vibrating agate ball-mill.

1.2 GAS SENSOR FABRICATION

Two typical structures for stannic oxide ceramic gas sensors are shown in Figures 1.1, 1.2, 1.3 and 1.4. The first is a directly heated type in which two heater coils are bonded inside the ceramic element itself. These separate heater coils also act as electrodes to detect changes in the conductance of the stannic

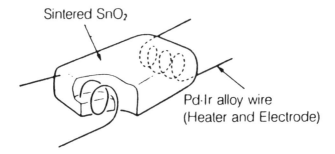

FIGURE 1.1   A gas sensing element of the directly heated stannic oxide ceramic type (e.g., the Figaro TGS 109).

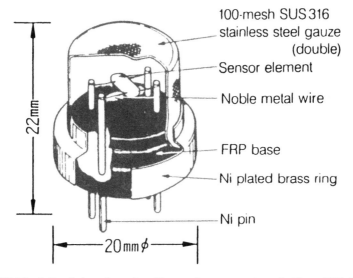

FIGURE 1.2   A directly heated stannic oxide ceramic gas sensor (e.g., the Figaro TGS 109).

oxide ceramic. The second is an indirectly heated type in which the single heater coil is separated from the ceramic by a tubular alumina substrate.

The two basic types are fabricated as follows.

a) *The directly heated type*

α-Alumina of 99.9% purity and 10 μm average crystal size is added to stannic oxide powder in a weight ratio of 1:1. The purpose of adding alumina is to provide ceramic strength, to modify the conductivity and to enhance the sensitivity.[4] Water is added and the aqueous paste is pressed into a die and sintered at a temperature above 700°C to produce a stannic oxide ceramic tip according to the sequence shown in Figure 1.5. This tip is mounted between two heater coils (of 90 μm diameter Ir-Pd alloy wire) previously welded to nickel pins on a plastic base. More paste is then formed around both heater coils, so producing

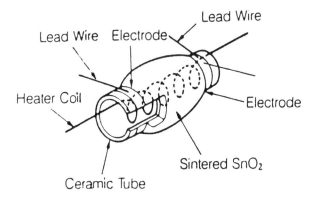

FIGURE 1.3 A gas sensing element of the indirectly heated stannic oxide ceramic type (e.g., the Figaro TGS 813).

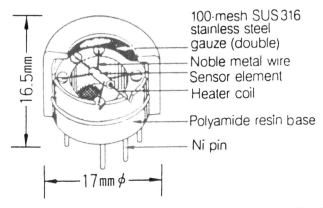

FIGURE 1.4 An indirectly heated stannic oxide ceramic gas sensor (e.g., the Figaro TGS 813).

a single sensing element. After air-drying, a binder is allowed to permeate into the element, which is again air-dried. Both heaters are then energized to raise the temperature of the ceramic to over 700°C for final sintering. The sensing element is then covered with a press-formed double 100-mesh stainless steel (SUS313) gauze cap. Fixed with a brass ring, the completed gas sensor shown in Figure 1.2 is obtained.

The double stainless steel gauze is an excellent flame arrester. It has been shown that when more than 500 sparks are produced inside the cap of a gas sensor installed in 2% isobutane, ignition never occurs in the surrounding isobutane-contaminated atmosphere.

b) *The indirectly heated type*

The substrate of this gas-sensing element is an alumina ceramic tube of outside diameter 1.2 mm, inside diameter 0.8 mm and length 8.6 mm, upon which two gold electrodes are paste-printed. Gold-palladium alloy leads are attached to each electrode using the same gold paste. These leads also act as stays to

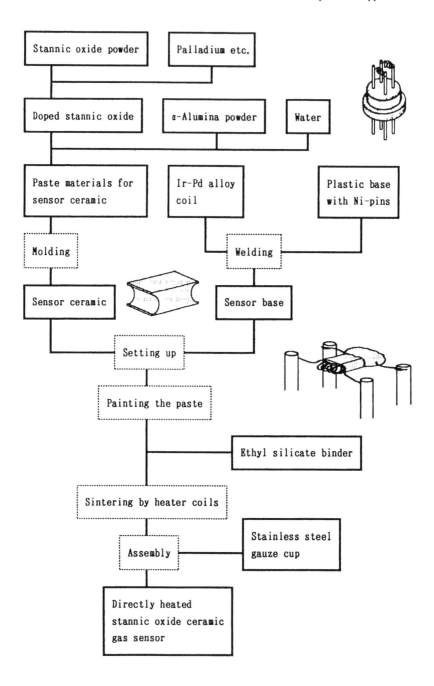

FIGURE 1.5   The manufacturing process of a directly heated stannic oxide ceramic gas sensor (e.g., the Figaro TGS 109).

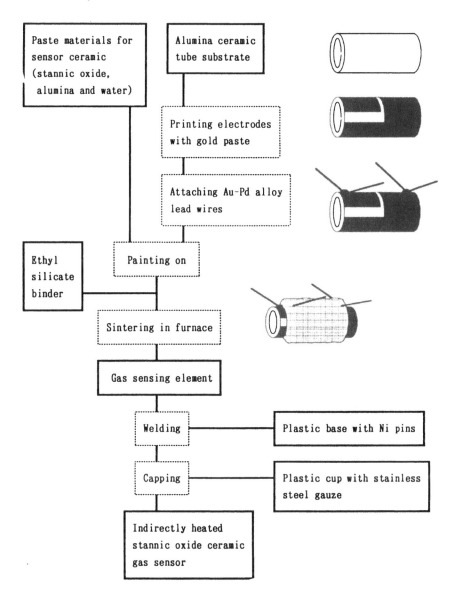

FIGURE 1.6 The manufacturing process of an indirectly heated stannic oxide ceramic gas sensor (e.g., the Figaro TGS 813).

support the substrate, and project from the electrodes as shown in Figure 1.6. In order to prevent cracking during sintering (due to the difference in thermal expansivity between the stannic oxide and the alumina substrate), the stannic oxide and the α-alumina (of average grain size 10 μm and 99.9% purity) are mixed in the weight ratio of 1:1, and 20 mg of the mixture is added to 20 mg of water, the resulting paste being painted over the two electrodes on the substrate to give about 0.2 mm thickness. After air-drying this unsintered element for 12 hours, the binder is allowed to permeate it, so strengthening the eventual sintered ceramic. The element is air-dried again for 6 hours and finally sintered at 725°C for 15 minutes in air using an electric furnace.

When the sintering is complete, an 80 Ω chromium alloy heater coil is inserted into the tubular substrate, and leads are welded to nickel pins mounted on a plastic base to finally obtain the sensor shown in Figure 1.4.

1.3   THE BINDER AND THE SINTERING PROCESS

An oligomer of tetraethyl orthosilicate is used as a binder in the stannic oxide ceramic gas sensor in order to enhance the sintering strength, as will be described. The amount of binder, converted to silica, which remains in the ceramic after sintering, is approximately 5.5% and this has the effect of improving not only the ceramic strength but also the characteristics of the sensor essential for practical use. The development of this binder has played a very important rôle in the successful commercial development of this form of sensor.

To appreciate how this comes about, the sintering process must first be understood in context. Initially, a powder is formed (by calcination) in which many spaces exist between the particles. Sintering should cause these particles to fuse to each other so that face and bulk mass transfer takes place with concomitant volume contraction and an increase in the strength of the material. Unfortunately, stannic oxide is a material which can only be poorly sintered, and it is shown below that neither changes in the calcining nor sintering parameters alone can improve the strength of the final ceramic significantly. This is why the binder is so important.

The effect of different calcining times has been investigated[5] as follows. Samples of stannic oxide powder were prepared using different calcining periods at 1200°C, then mixed with alumina powder and water to make paste for producing sensor ceramics. This paste was formed into shapes of 1.5 × 1.5 × 2.5 mm as test specimens, which were sintered at 725°C for 15 minutes, these being the same conditions as for actual sensor production. The volumes both before and after sintering were measured for each specimen, and are expressed in percentage form in Figure 1.7.

After sintering, a specimen of material calcined for a long period should show a greater volume contraction than that calcined for a short period. However, the results of these measurements do not suggest a clear correlation between calcination time and volume contraction for stannic oxide ceramic, and the percentages themselves are very low.

Similarly, Figure 1.8 shows volume contraction percentages after sintering for stannic oxide powders calcined at 450°C for 60 minutes, then ground in a vibrating agate ball-mill for different periods to obtain different grain sizes before sintering.[5] The results show that the stannic oxide with the smaller particle sizes does exhibit a greater volume contraction, but only by some 0.4%.

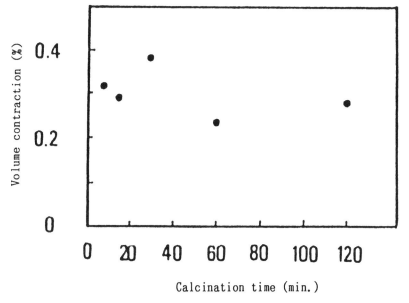

FIGURE 1.7 Relationship between calcination time to obtain stannic oxide powder and volume contraction of sensor ceramic caused by sintering. Calcination temperature: 1200°C; sintering temperature: 725°C; sintering time: 15 min.

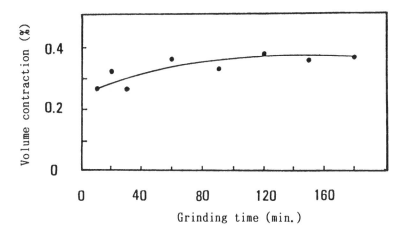

FIGURE 1.8 Relationship between grinding time of stannic oxide powder and volume contraction of sensor ceramic caused by sintering. Calcination temperature: 450°C; calcination time: 60 min; sintering temperature: 725°C; sintering time: 15 min; grinding: vibrating ball-mill.

Figure 1.9 shows the relationship between the shock resistance of actual stannic oxide ceramic gas sensors and the sintering or collapse strength of similarly prepared ceramic test specimens. The shock resistance was determined by observing the percentage of sensors cracked by applying a 400 G impact 10 times to each of 200 sensors, and the sintering strength by measuring

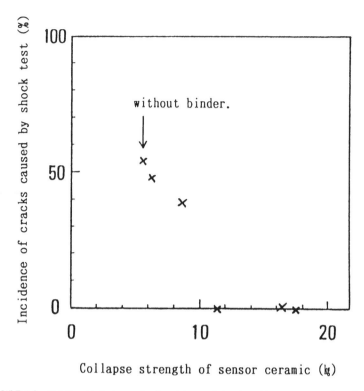

FIGURE 1.9   Relationship between shock resistance of stannic oxide gas sensor and mechanical strength of stannic oxide sensor ceramic. Number of tested gas sensors: 200 each; shock intensity: 400 G; number of shocks: 10; number of tested sensor ceramics: 5 samples each; sample size: 1.5 × 1.5 × 2.5 mm.

the mechanical loads needed to collapse 1.5 × 1.5 × 2.5 mm specimens of ceramic compressed vertically on the 1.5 × 2.5 mm sides. All sensors and associated samples were prepared using similar binder concentrations, and it will be seen that for those without binder (shown by the arrow in Figure 1.9) the collapse strength was low, the applied load being only 5.6 kg, whereas the cracked percentage was high. These data show that it is difficult to improve the strength of sintered stannic oxide by controlling the preparation parameters alone, and it is therefore fortunate that this can be achieved by using an appropriate binder. Commercially viable gas sensors must always use sufficient binder to give a collapse strength of more than 10 kg (as Figure 1.9 suggests), and as previously mentioned, this is a tetraethyl orthosilicate oligomer in the case of those manufactured by Figaro Engineering Inc.

The oligomer (a mixture of tetramer and pentamer, and abbreviated "ethyl silicate" hereafter) is generally used for refractories because it decomposes and generates a siloxane bond at high temperature and, during the process, forms a metallosiloxane bond with metal atoms of refractory material. (It is generally known that the oligomer acts effectively as a binder by forming a metallosiloxane-siloxane-metallosiloxane structure,[6] and the same effect would be expected for stannic oxide sintering.)

Table 1.1 shows the compounding ratio of ethyl silicate binder and its silica content along with the silica content of the eventual sensor ceramic; Figure 1.10 then graphs the collapse strength of this sensor ceramic as a function of its silica content. Here, it can be seen that with over 2.6% silica in the ceramic, a remarkable improvement in sintering strength is exhibited, and this conclusion confirms that the strength obtained by using ethyl silicate binder containing not less than 13% silica is sufficient for the fabrication of practical sensors.

The binder has other effects on the characteristics of gas sensors also; the major ones are

a)   The prevention of excess crystal growth
b)   An improvement in gas sensitivity
c)   An improvement in time-dependent performance

and each effect may be explained as follows.

a) *Prevention of excess crystal growth*

Figure 1.11 shows the change in crystallite size in stannic oxide as a function of sintering temperature, but with a fixed sintering time of 15 minutes. Sample sensors with 7% silica content were compared with those without binder. The crystallite size was calculated from a half-width of the X-ray diffraction peak of the (110) crystal face of stannic oxide ceramic peeled from sample sensors.[7] As shown in the figure, stannic oxide crystals tend to grow with increase in sintering temperature for samples without binder. However, the growth is very small for temperatures under 800°C when ethyl silicate binder is added. This implies that the silica produced during the decomposition of the ethyl silicate binder has the (known) effect of inhibiting crystal growth as the sintering process progresses.[8] Recalling the improving effect on sintering strength, it can be concluded that the ethyl silicate binder does not contribute to the sintering of stannic oxide itself via mass transfer on the surface, but permeates into the grain boundaries and decomposes into products with silica or siloxane bonds to prevent sintering between stannic oxide grains and so produces a network of bridges around the grains of stannic oxide and alumina which account for the increased strength.

The sizes of the stannic oxide crystals are closely related to sensor characteristics as described in the next section. So, in practical sensor production, the calcination process should be well-controlled to obtain uniform characteristics. It is, however, difficult to achieve good reproducibility when crystals grow quickly in the sintering process to form sensor ceramic, because uniformity in crystallite size obtained in the initial calcination process may be destroyed by this subsequent sintering. For this reason, the prevention of excess crystal growth is an extremely useful function of the ethyl silicate.

b) *Improvement in gas sensitivity*

Figure 1.12 shows the relationship between the sensor resistance and temperature in air, and in various gases. In this book, when not otherwise specified, "in

TABLE 1.1   Compounding Ratio of Ethyl Silicate Binder and Silica Content in the Sensor

| Sample number | Ethyl silicate binder | | | | Silica content in the binder (wt. %) | Silica content in the sensor $\dfrac{SiO_2}{SiO_2 + SnO_2 + Al_2O_3} \times 100$ |
| | Composition of the binder (vol. %) | | | | | |
	Ethyl silicate 40	Ethanol	Water	2% Hydrogen chloride		
1	80.0	13.0	6.0	1.0	32.0	7.0
2	65.0	27.0	7.0	1.0	26.0	5.5
3	32.5	50.5	16.0	1.0	13.0	2.6
4	16.0	58.0	25.0	1.0	6.4	1.3
5	—	—	—	—	—	0.0

Ethyl silicate 40: trade name of Japan Colcoat Chemistry Co. Ltd.

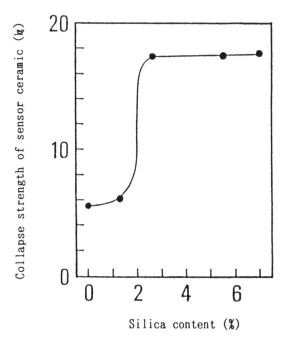

FIGURE 1.10 Relationship between collapse strength of stannic oxide sensor ceramic and its silica content. Sample size: $1.5 \times 1.5 \times 2.5$ mm; sintering temperature: 725°C; sintering time: 15 min.

air" means an atmosphere of normal air at an ambient temperature of 20°C and a relative humidity of 60%; "in gas" means an ambient atmosphere of normal air containing a certain concentration of that gas.

Samples both without binder and with 5.5% silica were tested, and it was shown that ethyl silicate binder results in the following modifications to sensor characteristics:

1. The absolute resistance falls (by about an order of magnitude).
2. The sensitivity to gases becomes greater, that is, the ratio of resistance in air to that in gases becomes greater.
3. The peak value of the resistance in air shifts towards a higher temperature.

The reason for the observed reduction in sensor resistance resulting from the inclusion of the binder is assumed to be that during sintering, as the ethyl silicate is decomposed into silica, oxygen is removed from the stannic oxide crystals, so increasing the number of oxygen defects and enhancing the n-type conductivity in the crystal bulk.

The improvement in response conferred by the binder is demonstrated for isobutane in Figure 1.13, where the normalized resistance change is plotted against the gas concentration for samples with different silica contents. Here,

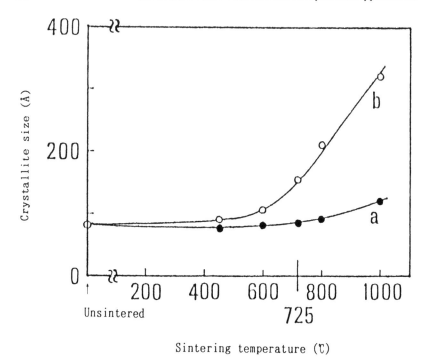

FIGURE 1.11 Influence of ethyl silicate binder on crystal growth of stannic oxide during sintering; (a): with binder (silica content 7%); (b): without binder; sintering time: 15 min.

the resistances are expressed as values relative to a reference resistance R_0 at 100 ppm isobutane. The relationship between this resistance ratio and gas concentration is almost a straight line on log-log graph paper as shown in the figure, and the gradient becomes steeper with increasing silica content derived from the ethyl silicate binder. This means that the presence of the binder makes the dependence of resistance on gas concentration greater — that is, the sensitivity is improved. Figure 1.13 also shows that 5.5% is the most suitable silica concentration for optimum sensitivity.

The peak value of sensor resistance in air (seen in Figure 1.12) is related to adsorbed oxygen on the stannic oxide surface, as will be described in detail in Chapter 2. So, it may be concluded that the reason this resistance peak in air shifts to higher temperatures when binder is present is that more adsorbed oxygen on the surface is maintained at a higher temperature by some interaction of the ethyl silicate and stannic oxide. Assuming that the oxygen adsorbed on the surface is greater around the temperatures at which gas sensing takes place, it is understandable (according to the gas sensing mechanism described in Chapter 2) that the sensitivity of the sensor to gas should also increase.

c) Improvement of time-dependent characteristics

Figure 1.14 demonstrates the time-dependent repeatability of the resistance in 2000 ppm isobutane for two groups of gas sensors, with binder, and without binder, respectively. The mean values for 50 sensors normalized by the initial

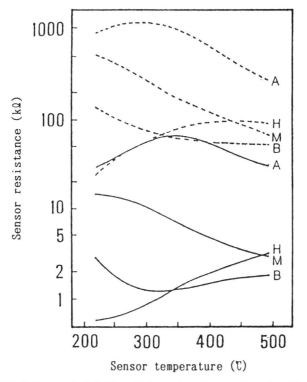

FIGURE 1.12 Influence of ethyl silicate binder on the resistance of stannic oxide ceramic gas sensors in gases at various temperatures. (———): with binder (silica content 5.5%); (------): without binder; A: air, H: hydrogen, M: methane, B: isobutane; gas concentration: 2000 ppm.

measurement value and its $\pm 3\sigma$ (where σ = standard deviation) are plotted. The resistances of samples without binder have a tendency to become higher and their distributions to become wider with time, as is shown in the upper graph. On the other hand, it is seen that the samples with binder do not show any such tendency and so exhibit excellent repeatability. It is assumed that the progress of sintering of the crystalline material in long-term operation is prevented by the ethyl silicate binder, resulting in these much-improved characteristics.

1.4 THE FINAL ACTIVE MATERIAL

Having introduced the basic features of stannic oxide gas sensor ceramic manufacture, the resulting material may now be considered in detail.

1.4.1 Relationship between Stannic Oxide Crystal Size and Sensor Characteristics

If it is assumed that the characteristics of the sensor ceramic are closely related to the unit stannic oxide crystal size, then in order to develop superior sensors

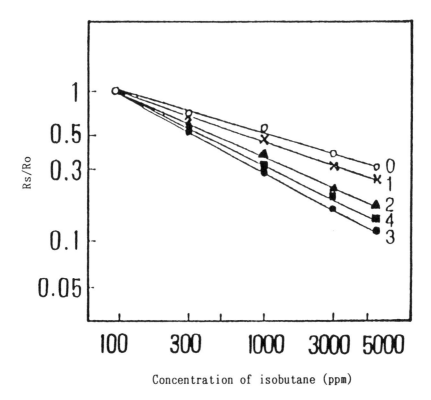

Concentration of isobutane (ppm)

FIGURE 1.13   Influence of ethyl silicate binder on the sensitivity of a stannic oxide ceramic gas sensor. R_o: sensor resistance in 100 ppm isobutane; R_s: sensor resistance in various isobutane concentrations; ceramic silica content: 0.0%, 1:1.3%, 2:2.6%, 3:5.5%, 4:7%.

and maintain high quality in mass-production, it is essential to understand this relationship and control the crystallite size appropriately.

As already described, stannic oxide powder as the raw material for the sensor ceramic is prepared by calcining tin hydroxide. Different calcination conditions produce different crystallite sizes, and as Figure 1.11 indicates, the original crystallite size in the stannic oxide powder itself is kept almost constant in the ceramic when sintered with tetraethyl silicate binder. Essentially then, control of the calcination conditions of the tin hydroxide makes it possible to control also the unit crystallite size of the stannic oxide ceramic, so that the characteristics of the eventual gas sensor become predictable.[10]

Figure 1.15 shows the powder method X-ray diffraction pattern of stannic oxide prepared by calcining tin hydroxide at 450°C for 1 hour. The diffraction peak for the (110) crystal face is the highest, and its ratio to the peaks of the reflex intensities for the other faces agrees with the pattern for the stannic oxide crystal in the tetragonal system. The pattern for stannic oxide calcined at 1200°C for 2 hours is the same, and shows that changing conditions influence crystal size through isotropic growth of crystal particles.

Figure 1.16 shows the change in crystallite size and surface area when the calcining time is varied at a fixed temperature of 1200°C. Here, the crystallite

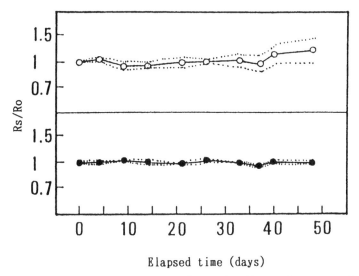

FIGURE 1.14   Change of resistance with time for stannic oxide ceramic gas sensors in 2000 ppm isobutane. R_o: the first measured resistance value; R_s: subsequent measured resistance values; (O—O): average of 50 sensors without binder; (●—●): average of 50 sensors with binder; (--------): ±3 times standard deviation of each average.

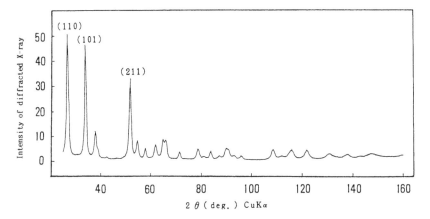

FIGURE 1.15   X-ray diffraction topograph of stannic oxide powder material.

size is estimated from the following formula, using a half-width of the (110) diffraction peak:

$$D_{110} = 0.9\lambda / \beta_{1/2}\cos\theta$$

(where D_{110} = crystal size, λ = 1.542 Å, $\beta_{1/2}$ = a half-width and θ = 13.3.°)[7] The data indicate that the crystal size becomes larger with increase in sintering time, whereas the crystal surface area becomes smaller.

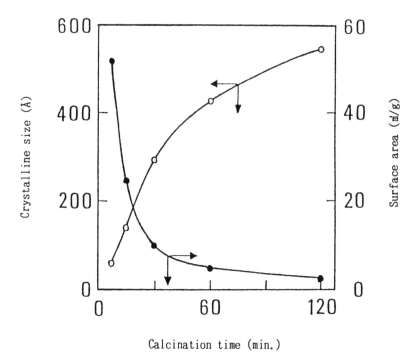

Calcination time (min.)

FIGURE 1.16   Relationship between calcination time for stannic oxide material and its crystallite size and surface area. Calcination temperature: 1200°C; (—O—) crystallite size; (—●—) surface area.

Figures 1.17 to 1.21 are scanning electron micrographs of stannic oxide powder material for various calcination times and show that lower order structural particles become larger as these calcination times increase. In Figure 1.17, several 3 to 4 μm particles look homogeneous, but it can be seen in Figure 1.21 that these particles were composed of approximately 0.5μm lower order structural particles. In Figure 1.17, such lower order structural particles are too small to be seen, but in Figure 1.21, they have grown large enough to be visible as a result of the longer calcination time. In accordance with the crystal growth curves shown in Figure 1.16, lower order structural particles grow and, hence, the structure of the higher order particles is changed.

Figures 1.22 and 1.23 show the relationship between sensor resistance and temperature in various gases for sensors using stannic oxide powder materials with different crystallite sizes and sintered at 725°C for 3 minutes. In Figure 1.22, sensor resistances are graphed both in normal air at 20°C and relative humidity 60% and in dry air at 20°C and below 100 ppm water content. Here, the curve of sensor resistance vs. sensor temperature in dry air shows a peak at approximately 400°C. This characteristic does not change its general shape in (comparatively humid) normal air, but the humidity reduces the actual resistance values because water vapor is essentially a gas. Such behavior is seen in all sensors, irrespective of crystallite sizes.

Similar curves, but for various reducing gases at concentrations of 2000 ppm in air, are also shown. Here, the sensor resistance markedly decreases, but

FIGURE 1.17 Scanning electron microphotograph of stannic oxide material. Calcination: 1200°C, 7 min.

resembles the curves for air to some degree, especially for methane and isobutane. The curve for hydrogen has no peak corresponding to those for air or other gases, which may be because hydrogen is a more powerful reducing agent and so rapidly consumes almost all the adsorbed oxygen so that the characteristic peak does not occur.

Another observation is that the sensor with the larger crystals has the higher resistance on the whole, as is demonstrated by Figure 1.24, which refers to various sensors operating at 450°C. However, when the resistance in air R_a is divided by the resistance in gas R_S, this is a measure of *sensitivity*, and Figure 1.26 shows that crystallite size does not have a very marked effect on this parameter.

The surface area of stannic oxide crystals decreases sharply and simply with crystallite size, as shown in Figure 1.16. However, the relationship between sensor resistance in gas and crystallite size shown in Figure 1.24 is not correspondingly simple. Sensor resistance in gas basically tends to increase with an enlargement of crystallite size, but the rate of increase progressively declines.

Figure 1.25 is useful in understanding this complex relationship between sensor resistance and crystallite size. It shows the relationship between the

FIGURE 1.18   Scanning electron microphotograph of stannic oxide material. Calcination: 1200°C, 15 min.

sintered density of stannic oxide ceramic, and the crystallite size of its powder material when that material is sintered to form a $1.5 \times 1.5 \times 2.5$ mm specimen under the same conditions as in normal sensor production. Here, it is seen that a smaller crystallite size gives a lower density, and the reason is thought to be as follows. Stannic oxide particles with high order structure, themselves composed of small crystals grown during the calcination process, have lower densities and are harder to pulverize because of the smallness of these crystals. So, the raw material powder with small crystals contains many low density particles. As a result, low density, high order structure is maintained even in the ceramic made from this small-crystal powder, and its bulk density is therefore correspondingly low.

It may be assumed that the rise in sensor resistance with crystallite size is caused by a decrease in oxygen defects in the lattices of the stannic oxide crystals during growth. The decline in the rate of resistance increase shown by sensors with larger crystallite sizes in Figure 1.24 is because some part of the fundamental resistance rise is offset by a conductivity increase with density, which follows from Figure 1.25.

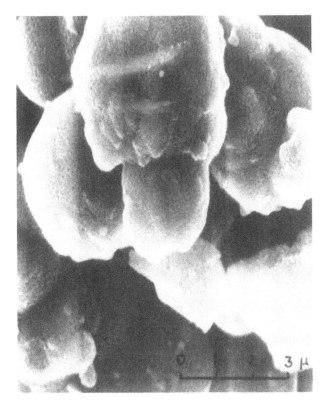

FIGURE 1.19 Scanning electron microphotograph of stannic oxide material. Calcination: 1200°C, 30 min.

At this point, it is useful to elucidate the concept of *gas sensitivity*. This is a common term which is defined as the ratio of sensor resistance in air to that in the relevant gas mix, R_a/R_S. This parameter obviously becomes larger as R_S becomes smaller, so that it does represent the *sensitivity* of a sensor, for, in general, R_S falls with increasing gas concentration. In the present context, gas sensitivity has been plotted in Figure 1.26 to compare sensors having different crystallite sizes operated at 450°C in 2000 ppm of various gases. Note that for isobutane and carbon monoxide, the sensitivity declines with increasing crystallite size; this implies that the change in resistance from that in clean air is smaller for sensors having larger crystals, whereas the converse is seen for methane. For hydrogen, high sensitivity is maintained over a wide range of crystallite sizes, but the apparent optimal value should be treated with caution, for the convex form of the curve is a result of many factors, and is not necessarily seen at temperatures other than 450°C.

The sensor resistance, both in normal air and in gas, changes in the same direction and to almost the same degree with increase in crystallite size, so these resistance changes offset each other and the effects of crystallite size on the gas sensitivity are actually rather small, as shown by Figure 1.26. However, the characteristics of each gas do affect the relationship between the sensitivity

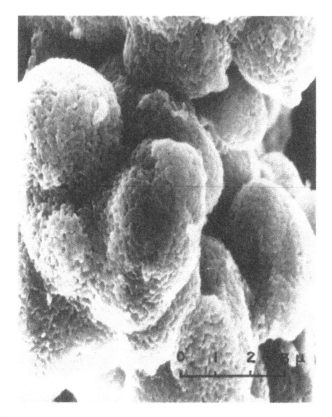

FIGURE 1.20   Scanning electron microphotograph of stannic oxide material. Calcination: 1200°C, 60 min.

and the crystallite size as is well demonstrated by hydrogen. A similar result is obtained for the correlation between gas sensitivity and the surface area of the stannic oxide powder, as shown by Figure 1.27.

Summarizing, it may be said that changes in crystallite size and surface area do not significantly affect sensitivity to general fuel gases, which is in contrast to the changes in the absolute values of the resistances involved.

From the viewpoint of practical usage, it is desirable that a sensor's resistance at a target concentration of gas in air should be several thousand ohms, for such values are very easily measured using simple electronic circuits. So, if the target gas is isobutane, for example, the crystallite size should certainly be under 250 Å, as implied by Figure 1.24.

1.4.2 The Sintered State of Stannic Oxide Ceramic

From the foregoing discussion, a reasonable "standard" crystallite size for a commercial stannic oxide ceramic gas sensor is about 250 Å, and to obtain this size, the calcining condition should be 450°C for 1 hour. After calcining, the stannic oxide material is ground into powder using a vibrating ball-mill to

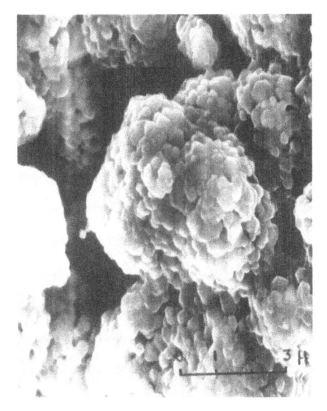

FIGURE 1.21 Scanning electron microphotograph of stannic oxide material. Calcination: 1200°C, 120 min.

obtain an average grain size of 3 μm (Figure 1.28). The distribution of grain size after grinding is shown in Figure 1.29, the data being based on sedimentation particle size analysis using an aqueous solution of 0.2% hexametaphosphoric acid as the dispersion medium.

The main optimum parameters relevant to the fabrication of standard stannic oxide material for gas sensors may now be listed:

1. Crystallite size (as determined by X-ray powder method): 250 Å
2. Average grain size (of higher order structure particles, as determined by the sedimentation method): 3 μm
3. Crystal surface area (as determined by the BET method): 18 m²/g

This stannic oxide powder material, mixed with α-alumina of 10 μm average grain size, is sintered at 725°C for 15 minutes to obtain stannic oxide ceramic, of which a scanning electron microphotograph is shown in Figure 1.30, where small stannic oxide particles are seen to fill the vacant spaces among the larger α-alumina particles. Compared with the stannic oxide powder immediately after grinding (Figure 1.31), the particles retain almost their original shapes and sizes. Melting due to sintering or fusion of the particles is

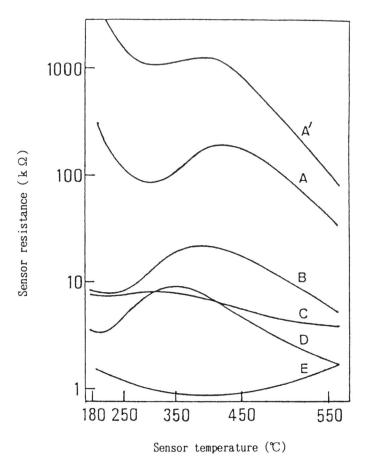

FIGURE 1.22   Relationship between the temperature of a stannic oxide ceramic gas sensor and its resistance in gas. Calcination of stannic oxide material: 1200°C, 15 min; average crystallite size: 59 Å; A: air, A': dry air, B: methane, C: carbon monoxide, D: isobutane, E: hydrogen; gas concentration: 2000 ppm each.

not observed, and the stannic oxide particles simply touch each other. This implies that particle contact and sintering strength are maintained mainly by the binder.

1.5   BASIC ELECTRICAL CHARACTERIZATION

In the preceding discourse, it has been tacitly assumed that the fundamental variable subject to measurement is the *resistance* of the sensor ceramic and that this may be measured either between the two heaters in a directly heated sensor, or across the two electrodes in an indirectly heated sensor. The fact that the resistance changes involved are considerable has been accommodated by employing logarithmic scales, as in Figures 1.12, 1.22, 1.23 and 1.24.

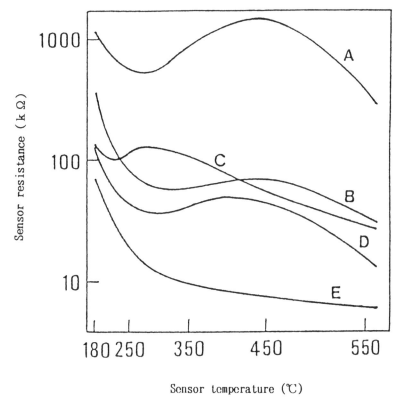

FIGURE 1.23 Relationship between the temperature of a stannic oxide ceramic gas sensor and its resistance in gas. Calcination of stannic oxide material: 1200°C, 60 min; average crystallite size: 428 Å; A: air, B: methane, C: carbon monoxide, D: isobutane, E: hydrogen; gas concentration: 2000 ppm each.

In addition, ratios of sensor resistance R_S to some datum resistance R_0 have also been graphed to indicate changes in an appropriate figure-of-merit. The implication here is that a datum sensor resistance R_0 is defined, and R_S is the sensor resistance under other conditions. This is a particularly useful approach where sensors having wide spreads of absolute values of resistance have to be compared, for it effectively normalizes the parameter, which may then be used in many different contexts, as will become evident in the text. For example, in Figure 1.13, the different slopes of R_S/R_0 clearly show how the greatest resistance change with increasing gas concentration can be achieved by optimizing the silica content of stannic oxide sensors.

The amount by which R_S/R_0 rises above, or falls below, unity indicates the fractional change in resistance relative to the chosen datum. This is well illustrated by Figure 1.32, which shows a set of typical response curves for a common Figaro sensor, the type 813. Here, it will be seen that the datum resistance R_0 is defined as the sensor resistance in 1000 ppm of methane, as shown by the heavy dot. Hence, all the values above this imply higher sensor

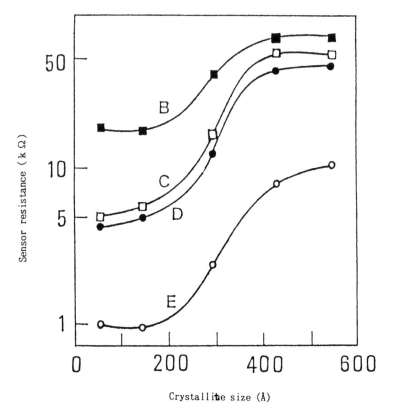

FIGURE 1.24 Relationship between the resistance of stannic oxide ceramic gas sensor in gas, and the crystallite size of the stannic oxide material: B: methane, C: carbon monoxide, D: isobutane, E: hydrogen; gas concentration: 2000 ppm each; sensor temperature: 450°C.

resistances, and those below it imply lower sensor resistances. The highest resistance ratio is, of course, for normal air, R_a/R_0; and the relative response to each gas is indicated by how much below this the relevant curve lies. This form of presentation is useful for comparing the responses of a sensor to various gases and also for comparing the responses of different sensors to similar gases, because the curves approach straight-line form on the log-log axes. Furthermore, it normalizes the responses of sensors having different values of R_0 so that average responses may be presented. However, it does not relate easily to the practicalities of circuit design for measurement purposes, and the resistance and conductance curves are more appropriate here,[12] as will be seen below.

As has been noted in Section 1.4.1, the *gas sensitivity* S_G of a sensor is commonly defined as the ratio of its resistance in air R_a to that in gas R_S. Figures 1.26 and 1.27 display this parameter as functions of crystallite size and surface area, respectively. This gas sensitivity can, in fact, be derived from resistance ratio diagrams like that of Figure 1.32:

$$S_G = R_a/R_S = R_a/R_0 \cdot R_0/R_S$$

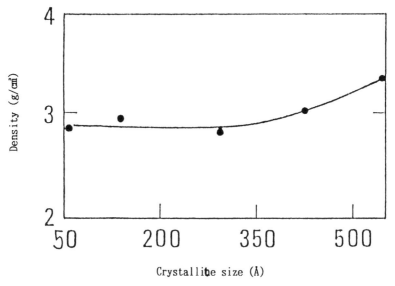

FIGURE 1.25 Relationship between density of stannic oxide ceramic and crystallite size of stannic oxide material. Sintering: 725°C, 3 min.

or

$$\log S_G = \log\{R_a/R_0\} - \log\{R_S/R_0\}$$

The first term here is the log of the *air level* (the normalized resistance of the sensor in air), and the second is the log of the normalized resistance under any other conditions. Thus, given a log ordinate, the vertical difference between the two must be the gas sensitivity.*

In general, the resistance of a sensor will fall with increasing gas concentration as sketched in Figure 1.33(a) on linear axes, where the rate at which this resistance falls is seen to diminish as the gas concentration increases. If the sensor resistance in air is R_a and its resistance in x% of gas is R_S, then the *chord sensitivity* AB is simply R_a/R_S. Notice that the *incremental sensitivity* — that is, the tangent to the curve — can be very different from this value, and varies considerably over the curve as a whole.

This resistance-vs.-concentration plot is reminiscent of a reciprocal curve, and indeed if the conductance G_S (= $1/R_S$) is also plotted on linear axes, it

---

Note: Gas sensitivity has been defined in this volume as

$$S_G = R_a/R_S = G_S/G_a$$

and its minimum value is obviously unity. In some papers, however, a slightly different definition appears:

$$S_G' = \Delta G_S/G_a = (G_S - G_a)/G_a = (G_S/G_a - 1)$$

so that the minimum value is zero, or $S_G = (S_G' + 1)$.

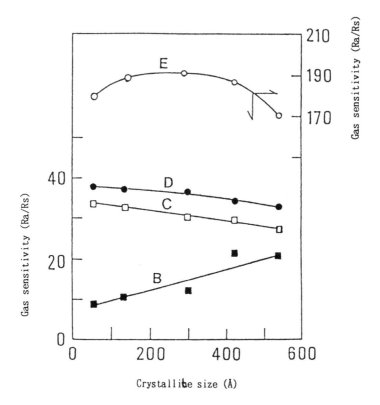

FIGURE 1.26   Relationship between the gas sensitivity of stannic oxide ceramic sensors and the crystallite size of the stannic oxide material. Gas sensitivity = (sensor resistance in air:R_a/sensor resistance in gas:R_s), B: methane, C: carbon monoxide, D: isobutane, E: hydrogen; gas concentration: 2000 ppm each; sensor temperature: 450°C.

appears in the general form of Figure 1.33(b). Here, at low gas concentrations, this conductance approaches a straight line, and the incremental sensitivity has a value similar to the chord sensitivity G_s/G_a in this region.

1.5.1   Volt-Ampere Characteristics

A bead of stannic oxide ceramic is fundamentally a negative-temperature-coefficient (NTC) thermistor, which means that its resistance in normal air will fall as its temperature rises. So, if this temperature (as defined by the heater) is first stabilized, and then a current is passed between the electrodes, further (Joule) heating can occur. Figure 1.34 shows this effect in the normal manner for thermistors, that is, on a volt-ampere graph. Here, the temperature of an indirectly heated sensor was raised by the heater to 450°C, and then increasing current was passed through the ceramic via the electrodes. As will be seen, the initial rise of current with voltage is at 45° on the log-log axes, which is a purely proportional, or ohmic, response. However, the curves then bend over, indicating that self-heating was taking place with a consequent fall in resistance. This effect even-

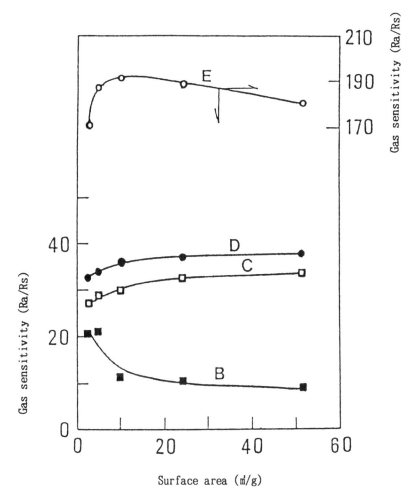

FIGURE 1.27 Relationship between the gas sensitivity of stannic oxide ceramic sensors and surface area of the stannic oxide powder material. Gas sensitivity = (sensor resistance in air:R_a/sensor resistance in gas:R_s), B: methane, C: carbon monoxide, D: isobutane, E: hydrogen; gas concentration: 2000 ppm each; sensor temperature: 450°C.

tually masks the resistance drop due to gas detection, rendering the sensor inaccurate. In fact, under extreme conditions, the sensor can be destroyed.

However, according to Figure 1.34, the problem can be avoided if the power dissipation due to self-heating is limited to some value $P_{S(max)}$. For a sensor resistance R_S and a sensor current I_S, this condition is:

$$I_S^2 R_S < P_{S(max)}$$

or

$$I_S < (P_{S(max)}/R_S)^{1/2} \tag{1.1a}$$

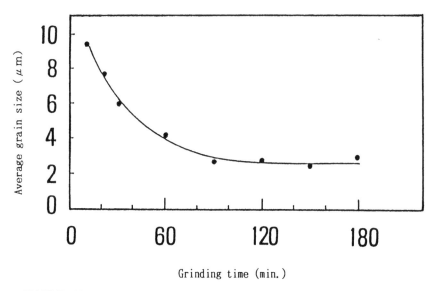

FIGURE 1.28   Relationship between grinding time of stannic oxide material and average grain size (weight average). Measurement: sedimentation method; dispersion medium: 0.2% hexametaphosphoric acid; calcination of stannic oxide material: 450°C, 1 h; grinding: vibrating ball-mill.

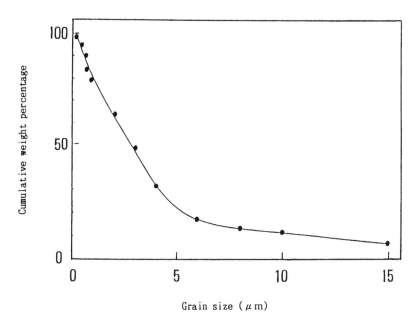

FIGURE 1.29   Cumulative weight percentage vs. grain size in stannic oxide powder. Measurement: sedimentation method; dispersion medium: 0.2% hexametaphosphoric acid; calcination of stannic oxide material: 450°C, 1 h; grinding: vibrating ball-mill 2 h.

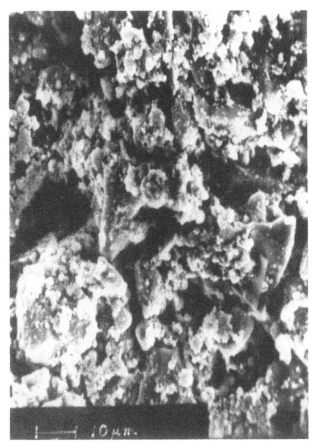

FIGURE 1.30   Scanning electron microphotograph of cross section of stannic oxide gas sensor ceramic.

For the Figaro-type material and geometry, this maximal dissipation is about 20 mW, so that

$$I_S < (0.02/R_S)^{1/2} \qquad (1.1b)$$

where I_S is in amperes and R_S is in ohms.

Sensor current limitation is best achieved by the associated circuitry, keeping in mind that the sensor voltage must not be reduced too far, otherwise linearity will be lost.

1.5.2 Polarity

The resistance of the stannic oxide gas sensor is, to a small extent, dependent upon the direction of the sensor current I_S; that is, it is slightly polarity-

FIGURE 1.31   Scanning electron microphotograph of cross section of stannic oxide powder material.

sensitive. This effect is never more than a few percent, and the offset is purely transient. Also, it may be obviated entirely by the use of alternating current.

Though the reason for this transient polarization remains obscure, it appears to be related to the material and structure of the electrodes and to the small amounts of characteristic-modifying additives which may be included.

Figure 1.35 serves to illustrate these points for a sensor in 2000 ppm of isobutane.[5] Firstly, a direct current I_S was passed through the sensor, which was allowed to stabilize; then the direction of I_S was reversed, as indicated in the figure. The resulting shift in resistance level is clearly seen, but recovers to its initial value within 25 days. This transient change is some 3.5%, but if measured using alternating current, the resistance stabilizes at about 2.6% below the DC level. Because such effects are so small, it may be concluded that surface phenomena are not involved, for they would be expected to lead to larger resistance variations.

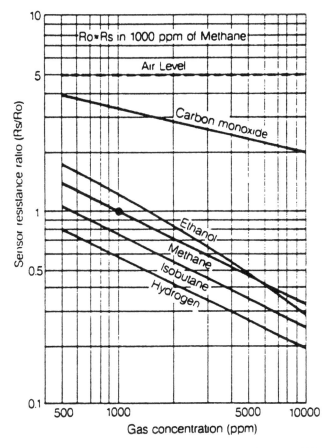

FIGURE 1.32   Typical resistance ratio characteristic for the Figaro type 813 sensor.

1.6   BASIC ELECTRICAL CIRCUITRY

Depending upon the type of sensor and its intended usage, the heater must maintain a stable temperature within the range 80 to 500°C. The requisite supply may be either AC using a transformer or DC using an integrated-circuit regulator. For most modern sensors, the heater voltage V_H is a nominal 5 V, though some are designed for 1 V to cover battery operation. (Specific examples of heater circuits will appear later in the volume).

As has been pointed out, the sensing current I_S through the active material must not give rise to further heating, and Figure 1.34 showed that for Figaro-type structures, this power should not exceed 20 mW. An extremely simple, though crude, method of providing both an output signal and a current-limiting facility, is to operate the sensor in series with a load resistor R_L, as shown in Figure 1.36. Here, a voltage source V_C is applied to the combination to provide the current I_S which drops voltages V_S and V_L across the sensor and load, respectively. This is a well-known circuit configuration in which the maximum

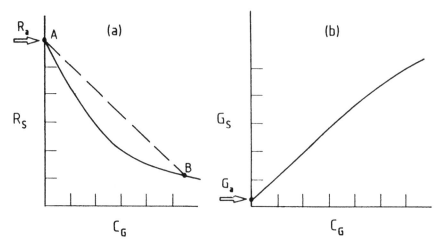

FIGURE 1.33   The general forms of (a) resistance and (b) conductance of the tin oxide gas sensor as functions of gas concentration, C_G.

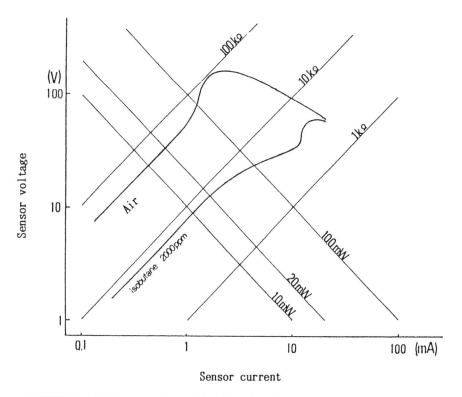

FIGURE 1.34   Volt-ampere characteristics of stannic oxide ceramic gas sensor.

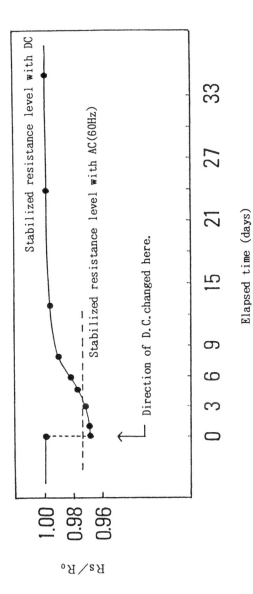

FIGURE 1.35 Resistance drift of stannic oxide ceramic gas sensor when the sensor current direction is reversed. Gas: 2000 ppm isobutane; R_o: stabilized resistance with DC; R_s: resistance at each measurement.

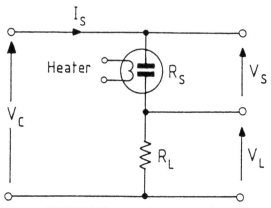

FIGURE 1.36 A simple sensor circuit.

power dissipation occurs when the resistances are equal, that is $R_S = R_L$ in this case. Hence, under these circumstances, each resistance dissipates half of the total power, which must be less than $P_{S(max)}$ for the sensor:

$$\frac{V_C^2}{4\,R_L} < P_{S(max)} \tag{1.2}$$

To illustrate this situation numerically in the case of Figaro sensors and to derive a maximum value for the applied voltage V_C, a commonly recommended value of 4 kΩ may be taken for R_L, whence

$$V_{C(max)} < (4\ R_L\ P_{S(max)})^{1/2}$$

or

$$V_{C(max)} < (4 \times 4000 \times 0.02)^{1/2} \simeq 17.9\ V$$

The output signal may be taken as a voltage across either the sensor or the load resistor, the latter being the most common:

$$V_L = \frac{V_C R_L}{R_L + R_S} = \frac{V_C}{1 + R_S/R_L} \tag{1.3}$$

If the actual sensor resistance is required, this can obviously be obtained from Equation 1.3:

$$R_S = R_L (V_C/V_L - 1) \tag{1.4}$$

Here, because V_C and R_L are known, it is necessary only to measure V_L.

Returning to Equation 1.3 again, it will be noticed that for low gas concentrations, when $R_S/R_L \gg 1$, and most of the available voltage is dropped across the sensor, its conductance becomes almost proportional to V_L:

$$V_L \simeq \frac{V_C R_L}{R_S} \text{ or } V_L \propto G_S \tag{1.5}$$

It has already been mentioned that at these low concentrations, G_S varies almost directly with the gas concentration C_G so that $V_L \propto C_G$ in these regions. Hence, it becomes reasonable to devise an electronic circuit which has an output proportional to G_S, and this is very easily done as shown in Figure 1.37.[13]

Here, an inverting operational amplifier configuration is used so that a virtual ground (or summing point) exists at the inverting input point. This means that the whole of V_C always appears across the sensor, and if the current into the amplifier itself is negligible, then,

$$I_S = V_C \, G_S = -I_F$$

Also, again because of the existence of the virtual ground point:

$$V_{out} = I_F \, R_F = -V_C \, G_S \, R_F \tag{1.6}$$

(Here, V_{out} is negative-going because of the inverting configuration.) That is

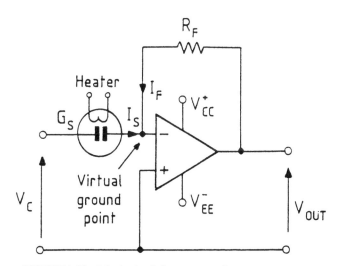

FIGURE 1.37 A basic circuit for sensor conductance measurement.

$$\left| V_{out} \right| \propto G_S$$

This is a very convenient way of measuring G_S, and scaling may be achieved by switching in various values of the feedback resistor R_F. Furthermore, the sensor conductance in clean air, G_a, may be backed off by simply extracting a small current from the virtual ground point using a negative supply (which must already exist to power the amplifier) and a preset resistor.

The power dissipated in the sensor is automatically limited by this circuit as follows.

$$P_S = V_C I_S = V_C \cdot \frac{V_{out}}{R_F} \qquad (1.7)$$

Numerically, knowing that typical operational amplifier power supplies are $\pm$ 15 V (so that the output voltage can never exceed this), and that for a Figaro-type sensor $P_{S(max)} < 20$ mW, suitable values for V_C and $R_{F(min)}$ may be determined:

$$\frac{P_{S(max)}}{V_{out}} = \frac{20}{15} \simeq 1.3 \geq \frac{V_C}{R_{F(min)}} \qquad (1.8)$$

where $R_{F(min)}$ is in kilohms.
A suitable pair of values would therefore be

$$V_C = 5 \text{ V and } R_{F(min)} = 3.9 \text{ k}\Omega$$

for example, both of which are common values in electronics. More detailed descriptions of circuits based on both techniques will appear at relevant points in the text.

REFERENCES

1. Iwasaki, I., et al., *Mukikagaku-zensyo XII-1-1* (Inorganic chemistry book), Maruzen, 1963, 194.
2. Samson, S. and Fonstad, C.G., Defect structure and electronic donor levels in stannic oxide crystals, *J. Appl. Phys.*, 44, 4618, 1973.
3. Ihokura, K., Tin oxide gas sensor for deoxydising gas, *New Mater. New Sci. Electrochem. Technol.*, 1, 43, 1981.
4. Taguchi, N., Japanese Patent, S50-30480.
5. Ihokura, K., Research on sintered tin dioxide gas sensors for combustible gases, Ph.D. thesis, Kyushu University, 1983.

6. Biddle, K.D. et al., The chemistry of ethyl silicate binders in refractory technology, *J. Appl. Chem. Biotechnol.*, 27, 565, 1977.

7. Nitta, N., *X-sen Kessho-gaku (X-ray Crystallography)*, 2nd ed., Maruzen, 1973, 512.

8. Matushita T., et al., Effects of additives on tin oxide-antimony oxide ceramics, *J. Ceramic Soc. Jpn.*, 80, 305, 1972.

9. Ihokura, K., Effects of tetraethyl silicate binder on a SnO_2 gas sensor, *Denki Kagaku (J. Electrochem. Soc. Jpn.)*, 51, 773, 1983.

10. Ihokura, K., The effects of crystallite size in sintered tin dioxide on changes of electrical conductivity in deoxydizable gases, *Denki Kagaku (J. Electrochem. Soc. Jpn.)*, 50, 99, 1982.

11. Seiyama, T., Gas sensors and catalysts, *Shokubai (Catalyst)*, 20, 80, 1978.

12. Watson, J., A note on the characterisation of solid-state gas sensors, *Sensors Actuators*, 8, 173, 1992.

13. Watson, J., The tin oxide gas sensor and its applications, *Sensors Actuators*, 5, 29, 1984.

{Data not specified are from K. Ihokura, Research on sintered tin dioxide gas sensors for combustible gases, Kyushu University, Ph.D. thesis (1983)}.

CHAPTER TWO

The Performance of the Stannic Oxide Ceramic Gas Sensor

2.1 CONDUCTIVITY VS. GAS CONCENTRATION CHARACTERISTICS

Having established the basic tenets of stannic oxide ceramic manufacture, it is now appropriate to consider further how the material performs in its gas sensing rôle. It has already been noted that the basic phenomenon is a decrease in resistance (i.e., increase in conductance) which occurs in the presence of combustible gases, and the electrical characteristics relevant to this phenomenon will now be introduced prior to an explanation of the actual mechanisms involved.

2.1.1 GAS CONCENTRATION CHARACTERISTICS

The increase in the conductivity of stannic oxide in the presence of a combustible gas is known to be associated with adsorption and desorption, and with redox reactions on the surface of the material,[1,2] and hence with gas concentration. Following this, the relationship between the conductance or resistance of a sensor and the gas concentration will be termed the *gas concentration characteristic*, while the term *gas sensitivity* will be reserved for the ratio R_a/R_S as defined in Sections 1.4.1 and 1.5.

A typical gas concentration characteristic for a stannic oxide ceramic gas sensor working at 450°C in isobutane is shown in Figure 2.1, and illustrates that the greatest incremental sensitivity occurs at the lower gas concentrations. This behavior is typical of all metal oxide sensors.[3]

A family of normalized gas concentration characteristics (using the resistance ratio R_S/R_0, where R_0 is the sensor resistance for 1000 ppm of isobutane) is shown in Figure 2.2 for five different gases. These plots are of straight line form on the logarithmic axes,[3] implying that the conductance of the active material may be approximated by the following expression:

$$G_S \ (= 1/R_S) \ = \ A \, C_S^{\alpha} \qquad (2.1)$$

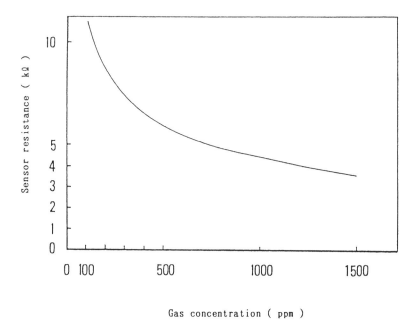

FIGURE 2.1 Typical gas concentration characteristic for a stannic oxide ceramic gas sensor. Gas: isobutane; sensor temperature: 450°C.

where G_S = sensor conductance, C_S = gas concentration, and A and α are constants.

At high gas concentrations, such plots deviate from their straight line forms because of temperature increases resulting from exothermic reactions on the surface. Figure 2.3 illustrates such a case. Here, a sensor temperature was first established at 450°C in clean air by operating its heater filament at constant power. An increasing concentration of hydrogen was then applied and the figure shows a flattening of the characteristic over about 11,000 ppm resulting from exothermic oxidation.

For sensors of different sizes, and hence different thermal capacities, the points where exothermic reactions become important will, of course, also be different. Furthermore, if different heater wire diameters are used, their different heat conduction capabilities will also influence this parameter.

2.1.2 Response Speeds

Figure 2.4 shows the response speed of a sensor which is suddenly inserted into, then removed from, atmospheres containing various concentrations of isobutane. At insertion, it is seen that equilibrium is reached more quickly for the higher concentrations than for the lower, which demonstrates a response speed dependency on gas concentration. This, in turn, implies that any chemical interactions (including chemisorption and redox reactions) must take place between the sensor and the gas in the atmosphere.

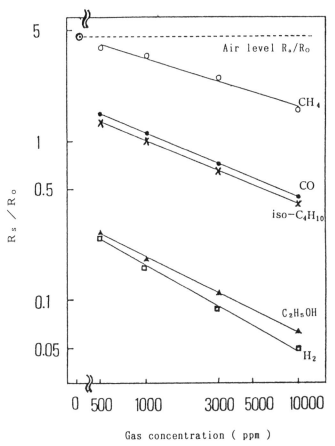

FIGURE 2.2 Normalized gas concentration characteristics for a stannic oxide ceramic gas sensor. R_0: sensor resistance in 1000 ppm isobutane (4.7 kΩ); R_s: sensor resistance in gases; sensor temperature: 450°C.

In contrast, the step decreases back to clean air show that the recoveries take place in similar times, implying no dependency on gas concentration. This is because recovery involves the readsorption of oxygen from the clean atmosphere to the previously oxygen-free surface, and since the oxygen partial pressure is constant, no recovery time variations would be expected.

2.2 SENSITIVITY VARIATION WITH TEMPERATURE

If it is assumed that the action of a sensor results from chemisorption and/or redox reactions on the surface, and knowing that the rates of such reactions are functions of temperature, then it would be expected that the surface temperature of the sensor would affect its observed characteristics considerably. Such dependencies will now be examined.

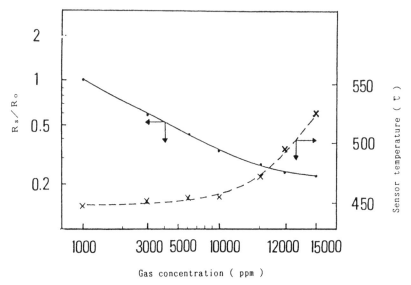

FIGURE 2.3 Normalized gas concentration characteristics for H_2 and sensor temperature change caused by H_2 combustion. R_o: sensor resistance in 1000 ppm hydrogen; R_s: sensor resistance at each gas concentration; (——●——): gas concentration characteristics; (----x----): sensor temperature.

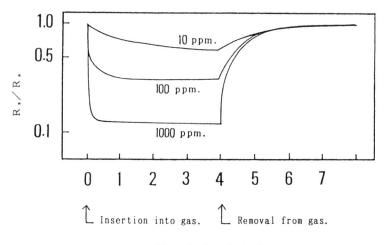

FIGURE 2.4 Response speeds of a stannic oxide ceramic gas sensor for various gas concentrations. Gas: isobutane; R_a: sensor resistance in air; R_s: sensor resistance in gas; sensor temperature: 450°C.

2.2.1 Sensor Temperature Characteristics

The term *sensor temperature characteristic* refers to a plot of the conductance or resistance of a sensor working in a given gas concentration, as a function of its temperature. Figure 2.5 shows a family of such plots for a series of gases,

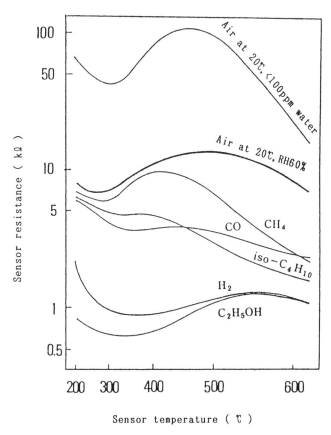

FIGURE 2.5   Sensor temperature characteristics of a stannic oxide ceramic gas sensor. Gas concentration: 1000 ppm.

each at a concentration of 1000 ppm, and also plots for both normal and dry air. Here, the sensor temperature has been varied by controlling the power supplied to the heater filament. Note that each curve exhibits a maximum and that this is most marked in the case of clean dry air (at 20°C and containing below 100 ppm of water).

Recalling that gas sensitivity is defined by the ratio R_a/R_s, note that the sensitivity to isobutane is greater than that to carbon monoxide at 600°C, whereas it is less at 350°C. Also, near 400°C, the sensitivity to methane is much lower than near 600°C. Such observations make clear the significant dependency of sensitivity on temperature, and Figure 2.6 illustrates this for a sensor in hydrogen and isobutane. These curves — like all such plots — are complex, but are useful in that for each gas, a temperature for maximum sensitivity can be clearly defined.

Usually, sensor temperature is maintained by a heater filament, as has been described. The power to such a heater is a function of the applied voltage V_H, and this leads to *supply voltage dependency*, which, in turn, suggests that V_H should be stabilized. (The heater power will not be quite proportional to the

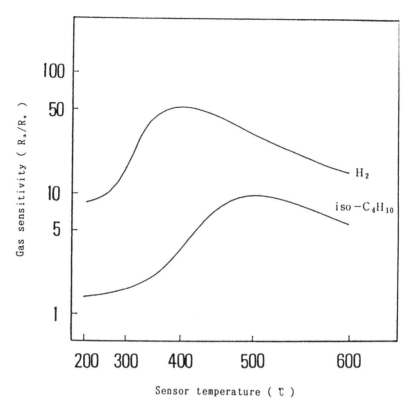

FIGURE 2.6   Relationship between gas sensitivity and sensor temperature for a stannic oxide ceramic gas sensor. R_a: sensor resistance in air; R_s: sensor resistance in gas; gas concentration: 1000 ppm.

square of the heater voltage, however, because of the change in filament resistance with temperature.)

2.2.2 Response Speeds — Further Observations

Whereas Figure 2.4 showed how the response speed varied as a function of gas concentration, but at a fixed temperature, Figure 2.7 illustrates how this response speed changes with temperature for a fixed gas concentration. For this latter series of measurements, the sensor was rapidly inserted into 1000 ppm of isobutane in air and allowed to reach equilibrium prior to rapid removal, its resistance during this procedure being plotted by permanent connection to a recorder.

The results indicate that at 350°C, the sensor hardly reaches equilibrium even after the elapse of 4 minutes, whereas at 500°C it requires only some 20 seconds. This difference is accounted for by the fact that the processes of adsorption and redox reaction are temperature-sensitive, as has already been noted.

Recovery to a high resistance level upon removal from the gas is also a function of temperature, and it will be seen that a marked "tail" appears in each case. This is because the recovery process involves both oxygen readsorption

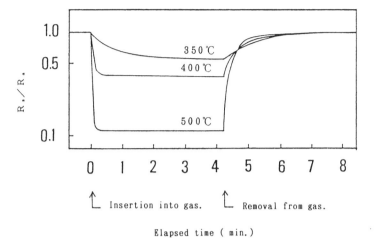

FIGURE 2.7 Response speeds of a stannic oxide ceramic gas sensor at various working temperatures. Gas concentration: 1000 ppm isobutane; R_a: sensor resistance in air; R_s: sensor resistance in gas.

at the surface and reoxidation of the stannic oxide. Reoxidation processes with high activation energies may well be the cause of this "tail".

2.3 THE EFFECTS OF OXYGEN PARTIAL PRESSURE AND HUMIDITY

Both the ambient oxygen partial pressure and the ambient humidity affect the sensitivity of stannic oxide ceramic sensors significantly, and though these effects are disadvantageous from a practical point-of-view, they do elucidate some of the mechanisms involved.

2.3.1 Oxygen Partial Pressure

Although the partial pressure of oxygen in ambient air is almost constant, it is useful to know how variations in this parameter affect the sensor resistance. Initially, the response to oxygen pressure *per se* may be obtained for a sensor at a constant operating temperature of 450°C by mounting it in a vacuum system and introducing oxygen so that a series of pressure levels is each held for some 20 to 30 minutes to permit sensor stabilization. The resulting resistance changes for a sensor working under these conditions are shown in Figure 2.8. Over much of this range of pressures, the resistance follows a straight line locus on log-log axes, so that,

$$G_S \; (= \; 1/R_S) \; = \; B \, P_{O_2}^{\,-\beta} \tag{2.2}$$

where G_s = sensor conductance, B and β are constants, and P_{O_2} is the oxygen partial pressure.

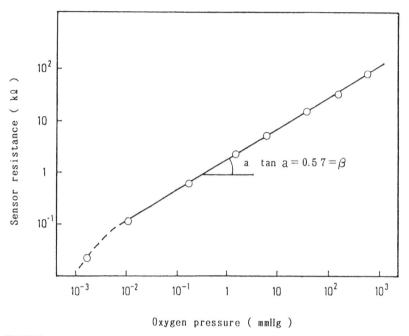

FIGURE 2.8 Relationship between the resistance of a stannic oxide ceramic gas sensor and oxygen pressure. Sensor temperature: 450°C.

A simplistic explanation for the fall in conductivity with oxygen pressure is that some oxygen is dissociated to form adsorbed O⁻ ions on the sensor surface, electrons being removed from the stannic oxide crystals. The decreasing concentration of electrons in the crystals would then result in decreasing conductivity. An analysis of this process shows that the conductivity of the material, and hence the conductance of a sensor, should be inversely proportional to the root of the oxygen partial pressure:[4]

$$G_s \propto P_{O_2}^{-0.5} \qquad\qquad (2.3)$$

According to this theory, the value of β in Equation 2.2 should be 0.5, but the slope of the characteristic of Figure 2.8 gives its actual value as 0.57. However, this value can actually be modified by including additives in the sensor material, or by changing its structure, such as the thickness of the active material. This is why the mechanism whereby the oxygen pressure controls the conductivity has been described as simplistic: it must actually be more complex in order that modifications to the value of β become possible at all.

In the region where Equation 2.3 is approximately true, the decrease in conductivity with increasing oxygen pressure is fully reversible. Furthermore, if the total atmospheric pressure is fixed by mixing oxygen and nitrogen, the same relationship applies with respect to the partial pressure of the oxygen.

Figure 2.9 gives normalized gas concentration characteristics in terms of R_S/R_0, where R_0 is the sensor resistance at 1000 ppm of isobutane in air at

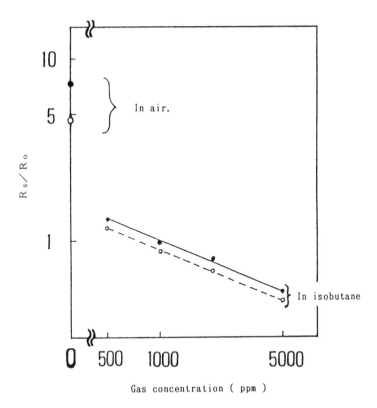

Gas concentration (ppm)

FIGURE 2.9 Normalized gas concentration characteristics at various oxygen partial pressures. Gas: isobutane; sensor temperature: 450°C; sample number: 60. (●———●): Air pressure: 1 atm (oxygen partial pressure: 0.2 atm). (○-------○): air pressure: 0.5 atm (oxygen partial pressure: 0.1 atm). R_O: sensor resistance for 1000 ppm isobutane at 1 atm; R_S: sensor resistance.

atmospheric pressure (760 mmHg) with an oxygen partial pressure of 0.2 atm. The upper characteristic gives the response to isobutane under these conditions whereas for the lower characteristic, the air pressure is 0.5 atm with an oxygen partial pressure of 0.1 atm. Furthermore, two points are shown which represent R_S/R_0 under the same two conditions but with no isobutane.

Notice that in the case of the clean atmospheres, the difference between the values of R_S/R_0 is larger than when isobutane is present. This implies that the fall in sensor resistance when the oxygen partial pressure is reduced is less when a reducing gas is present. This may be because the reducing gas continuously consumes oxygen adsorbed onto the surface of the stannic oxide, so lowering its concentration, as will be further explained in Section 2.5.3.

Although the proportion of oxygen in the atmosphere is sensibly constant, its partial pressure varies along with the barometric pressure, and such variations are reflected by stannic oxide ceramic sensors which are immersed in an atmosphere containing a constant gas concentration. This can give rise to an apparent drift in sensitivity which is sometimes erroneously ascribed to the sensor itself.

2.3.2 Water Vapor

The sensor temperature characteristics of Figure 2.5 show a marked fall in sensor resistance from dry air to humid air, and the gas concentration characteristic for water vapor given in Figure 2.10 confirms this. This latter plot was obtained by evaporating specific amounts of water into a chamber of known volume containing synthetic standard air (of humidity less than 10 ppm of water vapor) and maintained at a constant temperature. The sensor was run at 450°C inside this chamber, and the plot so produced indicates that water vapor may be treated as an electron donor gas, like combustible gases in this context. Yamazoe et al.[5] explain this phenomenon as being due to the formation of surface hydroxyl groups equilibrating with the water vapor as a function of its partial pressure.

For commercial stannic oxide gas sensors, it may be assumed that both combustible gases and water vapor will be present in the relevant atmosphere. Hence, it is desirable to know the effect of the water vapor on such sensors, and this may be elucidated by adding fixed concentrations of various gases to the test chamber described above. A family of normalized gas concentration curves (where R_0 is the sensor resistance for 2000 ppm of isobutane) is shown in Figure 2.11, from which it will be seen that water vapor can have a marked effect on the response, especially at low humidity levels. Over the gas selection utilized, the response to methane is modified by the greatest amount and that to isobutane by the least. A possible explanation for these variations in depen-

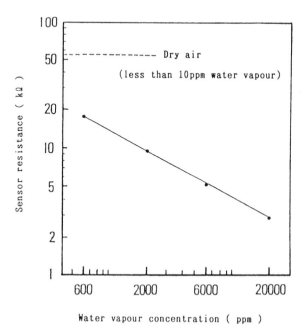

FIGURE 2.10   Gas concentration characteristic of a stannic oxide sensor in water vapor. Sensor temperature: 450°C.

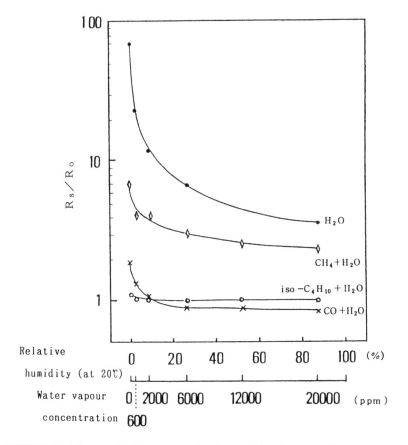

FIGURE 2.11 The normalized gas concentration characteristics of a sensor with respect to water vapor concentration in the presence of various gases. Gas concentration: 2000 ppm; atmospheric temperature: 20°C; R_O: sensor resistance for 2000 ppm isobutane and 12,000 ppm water vapor; R_S: sensor resistance; sensor temperature: 450°C.

dency may be that when both a combustible gas and water vapor react with the surface as electron donor gases at the same adsorption sites competitively, then the effect of the water vapor will be comparatively small, whereas when they operate at different sites, their effects will be cumulative, resulting in a marked humidity dependency.

Because of the practical importance of such changes in response resulting from the presence of water vapor, great attention must be paid to minimizing sensitivity to humidity in the development of commercially viable sensors.

A further — and longer term — effect of ambient humidity is illustrated in Figure 2.12. Here, a sensor is allowed to stabilize in standard air at 20°C and 60% relative humidity (RH) and containing 2000 ppm of methane. After 4 days, it is immersed in the same gas concentration, but in dry air, when its resistance is seen to rise rapidly, followed by a fall to below the initial stable value, this latter effect occupying some 10 days. Upon return to the initial conditions, a rapid fall in resistance is seen, followed by a slow rise to the initial

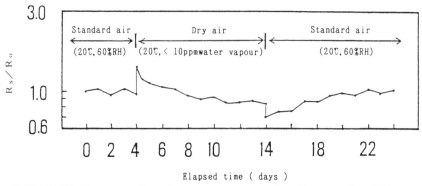

FIGURE 2.12   Long term effect of water vapor on a stannic oxide sensor. Gas: 2000 ppm methane; R_O: the first measured resistance; R_S: sensor resistance; sensor temperature: 450°C.

value. The implications of this are that when the humidity decreases, the sensitivity to methane appears to fall sharply at first, then slowly rises to an excessive value; and when the humidity increases again, it appears to rise sharply, then slowly recover to its initial value. Such long-term humidity-dependent effects are also seen in clean air when only humidity changes are involved.

Comparing the two phenomena of short- and long-term effects of water vapor, it may be that whereas the former is directly related to the oxidation of gas on the stannic oxide surface, the latter results from the formation of hydroxyl groups on the surface, but not accompanied by electron transfer.

2.4   THE EFFECTS OF VARIATIONS IN PREPARATION MODALITIES

The characteristics of production stannic oxide sensors are affected by several factors. For example, their resistance spreads are, in part, dependent upon variations in the areas of the electrodes deposited on the substrates. However, this section is primarily concerned with the effects of variations in the material preparation modalities, that is, the tin hydroxide calcination and subsequent sintering conditions.

2.4.1   The Effects of Calcining Conditions

In Section 1.4 it was shown that variations in calcining conditions produce differences in crystal sizes and surface areas, which, in turn, affect the resistances of sensors. However, their influence on sensitivity is not simple: it rises for some gases and falls for others, as is seen in Figure 1.26.

2.4.2   The Effects of Sintering Conditions

Stannic oxide ceramic is significantly affected by variations in sintering temperature and time, as was shown in Section 1.3, and Figure 2.13 extends this to demonstrate the influence of this temperature on sensor resistance.[6] Here,

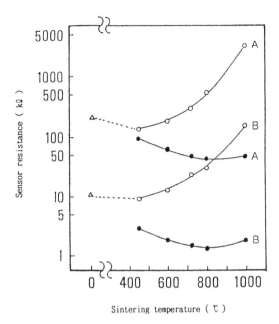

Sintering temperature (℃)

FIGURE 2.13 The effect of ethyl silicate binder on the resistance of the stannic oxide sensor. (●———●): With ethyl silicate binder (silica content 7%); (○———○): without binder; (Δ): dried in air, but not sintered; A: air; B: 2000 ppm isobutane; sensor temperature: 450°C.

two of the curves (open circles) show that R_s increases with sintering temperature for material without binder both in air and in 2000 ppm of isobutane.

This effect is somewhat difficult to explain. Figure 1.11 indicates that the crystallite size increases with temperature, so that the number of contact points should decrease, suggesting a rise in resistance, as is observed. However, the actual contact areas expand, which should lead to a fall in resistance, and this latter effect should be dominant. That it is not implies that another phenomenon must be occurring, and this may well be the decrease in oxygen lattice defects which results from high temperature annealing in air. These defects define the n-type semiconduction so that their reduction would be expected to result in a rise in resistance.

The curves in Figure 2.13 reflect the combination of these effects, the result of which is a net resistance increase with sintering temperature for material without binder. The effect on the gas sensitivity R_a/R_s is quite small, however, which can be deduced from Figure 1.26 knowing that an increase in the crystallite size results from an increase in sintering temperature.

In contrast, the curves with solid circles in Figure 2.13 represent sintering for material with binder and show that there is comparatively little change in sensor resistance with sintering temperature. This would be expected from Section 1.3, where it was shown that ethyl silicate binder has an inhibitory effect on the sintering process. Also, the binder removes oxygen from the crystal to produce silicon dioxide, and this reduction process increases oxygen electron-donor lattice defects which then contribute to n-type semiconduction.

As has been noted, Figure 1.11 shows that for a sintering temperature of 450°C, there is very little difference — only about 13% — in the crystallite size between stannic oxide with and without binder. This appears to mean that (in Figure 2.13) the condition of both ceramics sintered at 450°C should be similar in terms of crystallite size and degree of crystallinity regardless of the presence of the binder. Hence, the two sensors should also have similar electrical characteristics, including when gas is present. However, Figure 2.13 shows that a resistance difference of some 42% is exhibited. This supports the suggestion that, in the sintering process, the binder not only binds the crystals, but also reduces them, so increasing the number of electron-donor oxygen defects.

The curves with triangular points on Figure 2.13 refer to sensors which have been dried in air but not sintered. Though these sensors are mechanically weak, they nevertheless exhibit good gas detection properties, which is remarkable having regard to mechanisms which will be treated in Section 2.5.

2.5   THE MECHANISMS UNDERLYING SENSITIVITY

The amount by which the resistance of a stannic oxide sensor decreases with gas concentration in air is a function of several mechanisms, which will now be described.

2.5.1 The Surface Oxidation of Combustible Gases

It is well-known that in clean air, oxygen chemisorbs (that is, chemically adsorbs) onto the surface of a metal oxide catalyst and removes electrons from the conduction or valence band depending upon whether the oxide is n-type or p-type, so that its resistivity will either rise or fall. In the case of stannic oxide, which is an n-type semiconductor, electrons are removed from the conduction band, leading to a rise in resistivity. Then, when a combustible gas subsequently chemisorbs onto the surface, it is oxidized by the oxygen already there, so reinjecting the electrons and decreasing the resistivity as catalysis proceeds. An experiment to confirm this has been reported as follows.[15]

As shown in Figure 2.14, 9 25-mg stannic oxide sensor samples were prepared, but the heater filament of only one was connected to leads. These were loaded into an electric furnace (shown in Figure 2.15) which was run at 450°C, and helium carrier gas was allowed to flow through at a rate of 25 ml/min. The emergent gas was taken to a gas chromatograph, which made possible the quantification of any (noncarrier) gas released from the sample surfaces or of any other injected gas removed by these surfaces.

Initially, a series of 2 μl injections of isobutane was made until the gas chromatograph indicated that the full 2 μl emerged. This confirmed that none was retained by any reaction, implying that all of the adsorbed oxygen had been removed. Then, a series of 2 μl injections of oxygen was made at fixed intervals. At first, only about half of the injected oxygen was recovered via the gas chromatograph, showing that the rest had been adsorbed by the samples. However, after some 12 to 18 injections, the full amounts of oxygen were

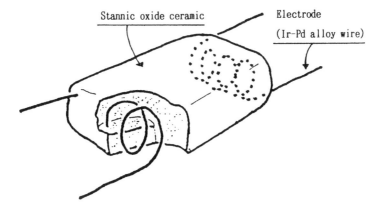

Stannic oxide ceramic

Electrode
(Ir-Pd alloy wire)

FIGURE 2.14 Stannic oxide ceramic sample for resistance measurement.

recovered, showing that all the active adsorption sites on the samples had adsorbed oxygen. At this point the measured resistance of the test sample was found to be high, as expected.

A series of 2 µl injections of isobutane was then made, and the gas chromatograph at first recorded the presence of carbon dioxide, water vapor, and a small proportion of isobutane. After seven or eight injections, the whole of the isobutane in each injection was recorded, showing that all of the previously adsorbed oxygen had been removed. The resistance of the test sample was then found to be low, again as expected.

When another series of 2 µl oxygen injections was made, the resistance of the test sample recovered to its high value again.

Figure 2.16 and Table 2.1 illustrate this sequencing. Figure 2.16 begins at the left with the tenth of a series of isobutane injections and indicates that all of the isobutane was recovered. Then 15 oxygen injections were made, and the chart shows that the full 2 µl was recovered at the 10th injection. Table 2.1 gives the resistance of the test sample as 26.21 kΩ at the 13th oxygen injection.

At this point 10 injections of isobutane were made and Figure 2.16 shows that at the 8th, the full 2 µl was recovered. Table 2.1 shows that the test sample resistance had fallen to 460 Ω by the 8th injection. It also shows that this resistance returned to a high value (25.87 kΩ) after a further 13 injections of oxygen.

These results suggest that:

1. The high resistance condition of the ceramic depends upon adsorbed oxygen, as is confirmed by its variation with oxygen partial pressure illustrated in Figure 2.8.
2. The catalytic oxidation of isobutane occurs at the ceramic surface, leading to the removal of adsorbed oxygen and also lattice oxygen in the surface layers.
3. The resistance of the ceramic decreases as its adsorbed oxygen is removed.

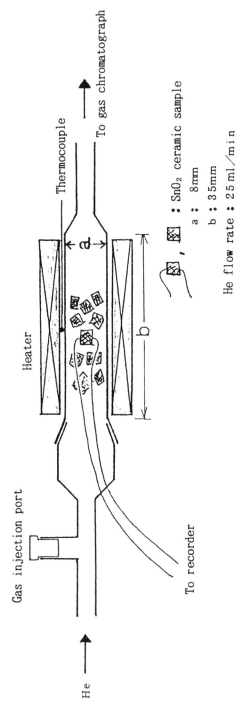

FIGURE 2.15 Equipment for examining the relation between gas reaction and resistance of stannic oxide ceramic samples.

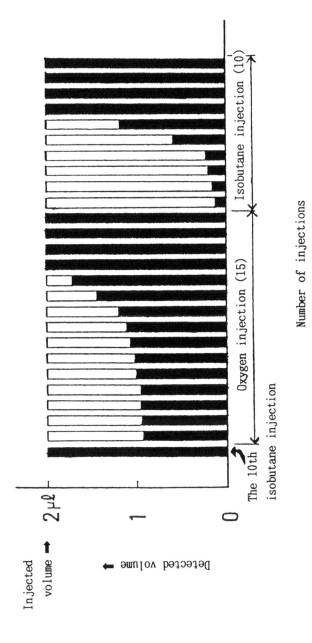

FIGURE 2.16 Detected volume variation for injected gas by gas chromatograph. Carrier gas: helium; detector: TCD; column: molecular sieve 5 Å, 80°C.

TABLE 2.1   The Resistance of an Experimental Sensor Sample in Helium after the Application of Oxygen and Isobutane

	After injecting with 26 $\mu\ell$ oxygen	After injecting with 16 $\mu\ell$ isobutane	After reinjecting with 26 $\mu\ell$ oxygen
Resistance of SnO$_2$ ceramic (kΩ)	26.21	0.46	25.87

Although the experimental conditions were not quite the same as those for a normally operating sensor, they did demonstrate that the mechanisms underlying the sensitivity to gas are directly related to the adsorption of oxygen and the subsequent oxidation of combustible gas using this adsorbed oxygen. This catalytic activity (for both isobutane and propylene) had been confirmed by Seiyama and others.[10,11]

2.5.2 Adsorbed Oxygen on the Sensor Surface

It is known[7] that the oxygen chemisorbed onto the surface of stannic oxide is in the form of three negatively charged ionosorbed species, O_2^-, O^- and O^{2-}. Of these, O_2^- is formed at about 100°C, while the others appear at the much higher temperature of about 400°C and remain on the stannic oxide surface up to 520°C according to Yamazoe et al.[5]

Using Figaro stannic oxide ceramic gas sensors and employing temperature programmed desorption (TPD) and electron spin resonance (ESR) methods, it has been confirmed[8] that O_2^- exists on the surface only in the temperature range below 150°C, while O^- (specified using ESR) remains up to some 400°C. Figure 2.17 illustrates the results of 5 tests on such sensors using TPD and shows that some oxygen is still being desorbed even above 400°C. However, between 400 and 500°C, though TPD methods confirmed this desorption, ESR methods did not register signals for O^- (and are not sensitive to O^{2-}), so it is possible that most of the desorbed species was O^{2-}, which is loosely bonded. Above about 600°C, only lattice oxygen should be desorbed.

The relationship between sensor resistance and adsorbed oxygen has been investigated by Egashira et al.[9] with the results shown in Figure 2.18(A). Here, the resistance is seen to decrease (right to left) for progressively lower amounts of adsorbed oxygen, at first rapidly, then more slowly. Initially, the higher rate of resistance decrease is taken as resulting from the desorption of O^- and O^{2-}, which are therefore assumed to play a major role in defining sensitivity at these levels.

Figure 2.18(B) is from a graph by Iwamoto et al.[8] relating sensor resistance to the volume of adsorbed oxygen remaining after desorption at a series of temperatures. Desorption which takes place at temperatures from 200 to 550°C results in a large decrease in sensor resistance as can be clearly seen. It is largely O^- which is desorbed at these temperatures (as demonstrated by the ESR work), which suggests that it is mainly this anion which remains adsorbed on the surface and which therefore defines the high resistances of sensors operated near 400°C. This supports the dependency of stannic oxide resistance on oxygen partial pressure suggested by Figure 2.8.

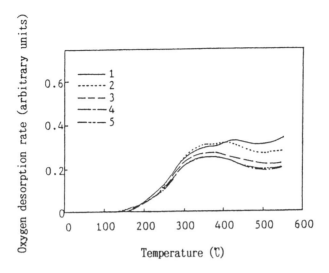

FIGURE 2.17 Temperature programmed desorption of oxygen from the stannic oxide gas sensor. Sample: Figaro gas sensor including 2% ethyl silicate binder. (Numbers indicate a series of five tests.) (Iwamoto, M., *Proc. 34th Meet. Jpn. Assoc. Chem. Sensors, Electrochem. Soc. Jpn.* August 1989, 23.)

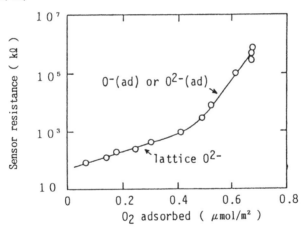

FIGURE 2.18(A) Relation between SnO_2 sensor resistance at 200°C and the amount of adsorbed oxygen. Preadsorption of O_2 was by 800°C → R.T. (From Egashira, M., Proc. Symp. Chem. Sensors, 1987 Joint Congress, Electrochem. Soc. Japan and U.S., Honolulu, October 1987, 39.)

Because the area of stannic oxide surface occupied by adsorbed oxygen is, at most, some 2%,[5,8] the implication is that chemisorption takes place only on those defects which make the phenomenon possible — if all defects led to chemisorption, the proportion of covered area would be much greater.

To recapitulate, though the resistance of an n-type semiconductor decreases monotonically with increasing temperature, Figure 2.5 shows that the resistance of a stannic oxide sensor does not behave in this way, which suggests that

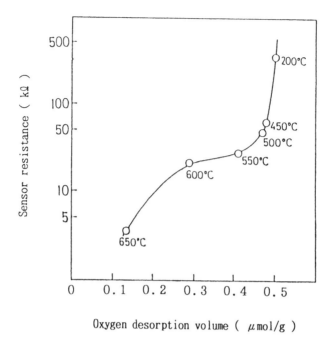

FIGURE 2.18(B)   Relationship between SnO$_2$ sensor resistance and remaining adsorbed oxygen on SnO$_2$. Preadsorption of O$_2$ was by 800°C → R.T.; oxygen desorption was carried out at each temperature indicated in the figure; sensor resistance was measured at 200°C in He. (From Iwamoto, M., *Proc. 34th Meet. Jpn. Assoc. Chem. Sensors, Electrochem. Soc. Jpn.* August 1989, 23.)

other factors are involved, and these have now been seen to include the effects of chemisorbed oxygen anions.

Also in Figure 2.5, the curve for dry air has a marked local maximum at about 470°C, and this may now be explained as follows. It has been seen that when stannic oxide is heated in air, the negative ion species O$^-$ and O^{2-} are formed at about 400°C with a concomitant rise in resistance. However, the formation temperature can differ from the 400°C region as a result of variations in the preparation of the stannic oxide. In the present case, if it is assumed that the O$^-$ and O^{2-} species are formed at about 470°C, this would account for the increase in resistance up to the local peak value for dry air seen in Figure 2.5. Above this temperature, these species will begin to desorb, and this will result in a fall in resistance, as is also observed.

Summarizing, the initial fall in resistance at low temperatures is due to the normal bulk semiconductor negative temperature coefficient (NTC); the rise to a local peak is due to the formation of the O$^-$ and O^{2-} species; and the subsequent decrease in resistance at high temperatures is due to the combination of both the desorption of these species and the NTC of the bulk material.

2.5.3 Gas Sensitivity

Simplistically, it might be thought that the fall in resistance of a stannic oxide ceramic sensor in the presence of combustible gas could be explained as follows. Initially, the resistance is high because the oxygen ion species on the surface have been negatively charged by receiving electrons from the crystal. Then, when a combustible gas comes into contact with these adsorbed oxygen species, it reacts with them to form such products as CO_2 and H_2O.[10,11] It could then be assumed that electrons are returned to the crystal surfaces and grain boundaries and, hence, decrease the resistance.

Figure 2.9 shows that the gas concentration characteristic is not greatly affected even when the oxygen partial pressure is halved, which implies that a simple resistance decrease with increase in bulk carrier density by adsorbed cations is not the only mechanism at work. Figure 2.19 illustrates an alternative mechanism which involves a reaction cycle consisting of the catalytic oxidation of a gas (carbon monoxide in this illustrative case) on the sensor surface resulting in the consumption of adsorbed surface oxygen. This is followed by replacement of the oxygen from the surrounding air.[12] The equilibrium density of the adsorbed surface oxygen with the concentration of ambient gas would then define the sensor resistance.

Returning to Figure 2.6, it is seen that the gas sensitivity vs. temperature curve for isobutane is different from that for hydrogen by virtue of being both lower in magnitude and shifted to the right — that is, to the higher temperature end of the scale. From the foregoing paragraphs, these differences could be interpreted as being due to the different reactivity of each gas to the negatively charged adsorbed surface oxygen and also the differences in the actual reaction

$$CO + \bar{O}ad\,(SnO_{2-x}) \longrightarrow CO_2 + (SnO_{2-x})^*$$

$$1/2\,O_2 + (SnO_{2-x})^* \longrightarrow \bar{O}ad\,(SnO_{2-x})$$

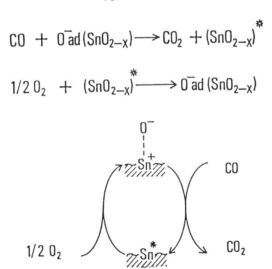

FIGURE 2.19   Possible scheme of CO oxidation on SnO_{2-x}.

rates which would result from different densities of this oxygen at different temperatures. Differences in reaction rates would also depend on the gas concentration, and this is reflected in the times to reach a steady state depicted in Figure 2.4.

2.5.4 The Effect of the Crystal Surface on Gas Sensitivity

As seen above, the equilibrium density of adsorbed oxygen on the stannic oxide surface is, in part, dependent upon the ambient concentration of combustible gas, and it is this equilibrium density which determines the sensor resistance in gas.

If the gas sensitivity was affected by the surface charge carrier concentration alone (this being defined by the simple exchange of electrons between the stannic oxide surface and the oxygen on it) then this sensitivity should vary with the specific surface area. However, Figure 1.27 does not show such a relationship — in fact, for gases such as methane and hydrogen, the sensitivity actually falls as the surface area increases. Although the experiment relating to this figure did not address the question of whether any changes other than those in surface area were involved, the results do imply that other mechanisms must be in operation. That is, explanations involving changes in carrier concentration alone are again seen to be too simplistic.

It should be noted here that the sensor temperature characteristics given in Figures 1.22 and 1.23 differ considerably for samples having different surface conditions. For these samples, the calcination conditions were different, which should have led to different surface defect conditions, as implied by the photomicrographs of Figures 1.18 and 1.20. The adsorpion states of the surface oxygen must, therefore, also be different, and this situation is reflected by the different positions of the local maxima in air.

These different oxygen adsorption states result in a variety of sensitivities to combustible gases, but the relationship is not always simple and is different for each gas. This too is a result of the differing reactivities of the various surface oxygen states to each gas. A concomitant of this is that changes in surface condition resulting from different calcination conditions alter and define the relative sensitivities to various gases.

2.5.5 The Mechanism of Conductance Change

Changes in sensor resistance, even with low gas concentrations such as 1000 ppm, are large, giving rise to high sensitivity, as is evidenced by Figures 2.2 and 2.5. However, not only have surface area effects been shown above not to modify sensitivity significantly, but it has been pointed out (in Section 2.5.2) that only some 2% of this surface is covered by adsorbed oxygen. Hence, the observed large changes in resistance cannot be ascribed only to carrier density variations in this small fraction of the total area. Noting that the ceramic is actually an aggregate of many high order structural particles, as shown in Figure 1.30, it is therefore likely that effects involving the vast number of grain boundaries must play a major role.

The stannic oxide crystal is fundamentally an n-type semiconductor containing oxygen lattice defects which act as electron donors.[1] When oxygen atoms remove electrons from the donors near the surface to become adsorbed as negative ions, a space charge layer is formed, along with a concomitant potential barrier, the magnitude of which varies with the concentration of adsorbed oxygen. The relevant energy band diagram is given in Figure 2.20, and the concept may be extended to give a potential barrier diagram between grain boundaries as shown in Figure 2.21.

The depth of the space charge layer is the Debye length, and when the crystallite size is larger than twice this, the potential barrier at the grain boundary becomes a major factor in defining the sensor resistance. Thus, when the density of the negative adsorbed oxygen ions is large, the potential barrier is high, resulting in a high resistance; when the density is reduced by the presence of combustible gas, the potential barrier falls, as does the resistance. This mechanism is illustrated in Figure 2.22(A).

Conversely, when the crystallite size is smaller than twice the Debye length, each crystal acts like a channel in a field-effect transistor (FET), and the negative surface charge due to the adsorbed negative oxygen ions has an effect like that of a FET gate voltage and affects the potential energy level E_C even inside the crystal. Figure 2.22(B) illustrates this situation, and it will be seen that for clean air, the grain boundary potential never falls to the E_C level, so that the resistance is very high. For an atmosphere containing a combustible gas

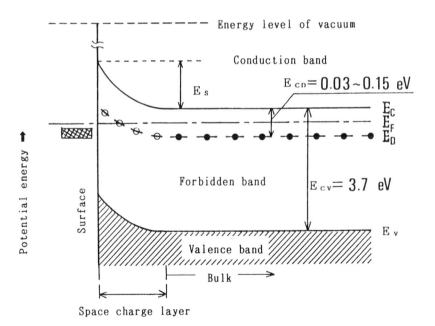

FIGURE 2.20 Band diagram for SnO_{2-x} with negatively charged adsorbed oxygen. (■): Surface potential by adsorbed O_2^- or O^-. ($O_2 + e \Leftrightarrow O_{2\,ad}^-$, or $1/2\,O_2 + e \Leftrightarrow O_{2\,ad}^-$). E_s: potential barrier; E_F: Fermi level; E_D: donor level; E_C: lowest level of conduction band; E_V: highest level of valence band; E_{CD}: depth of donor level; E_{CV}: energy gap between E_C and E_D.

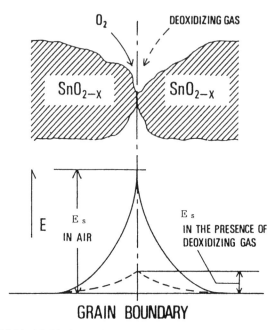

FIGURE 2.21   Model of potential barrier at grain boundary. E_s: potential barrier.

(where the adsorbed oxygen density falls), the grain boundary potential is reduced, leading to a lower resistance, but it does not fall to the E_C level, as is also shown in the figure.

These mechanisms account for the very large observed fall in sensor resistance when combustible gas is encountered, compared with the value in clean air.

Early reports on the value of the Debye length for stannic oxide have not proved very useful in explaining the sensitivity mechanism satisfactorily. However, in the case of zinc oxide, the Debye length has been given as between 3000 Å at 27°C and 4000 Å at 550°C;[13] and if stannic oxide exhibits similar values, then this, taken along with a typical crystallite size of 250 Å (as described in Section 1.4), supports the "FET mechanism". However, because the stannic oxide ceramic is composed also of 2 to 4 μm higher order crystal agglomerations or particles which contact loosely with each other (as implied by Figure 1.29), then the former mechanism might be operative, too.

2.6   TRANSIENT BEHAVIOUR

When a stannic oxide sensor is energized after it has been stored in ambient atmosphere over a long period, it is subjected to a temperature rise from room temperature to several hundred degrees centigrade. Two transients having different time constants are then observed prior to final stabilization, of which the first is a short-term transient lasting several minutes and usually called the *initial action transient*. The other is the *long-term transient* which may last

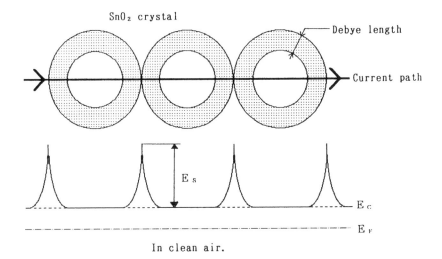

FIGURE 2.22(A)   Concept of potential along current path in SnO_2 sensor (crystal diameter > twice Debye length). E_S: potential barrier; E_F: Fermi level; E_C: lowest level of conduction band; ( ): space charge layer.

from a week to a month depending upon the type of sensor involved. These transients will now be explained separately.

2.6.1 The Initial Action Transient

The short-term transient is characterized by a fall in the sensor resistance immediately after energization, followed by recovery to a higher stable value within a few minutes, as shown in Figure 2.23.[20] The transition period required to reach this initial stable resistance level (usually called the *initial stabilization time*) depends not only on the type of sensor, but also upon its recent history. In general, the longer the sensor has been stored, the longer will be the initial stabilization time. However, this initial stabilization time does reach a plateau for very long storage times, as shown in Figure 2.24.[20]

Knowing that the initial action transient does not occur in vacuum, and that the sensor resistance in air depends upon the adsorption of oxygen, the transient behavior may be explained as follows.

When the sensor is energized, its temperature rises to an operating level (such as 450°C) within 20 to 40 seconds. Initially, the resistances of the constituent stannic oxide crystals decrease rapidly by normal semiconductor action involving

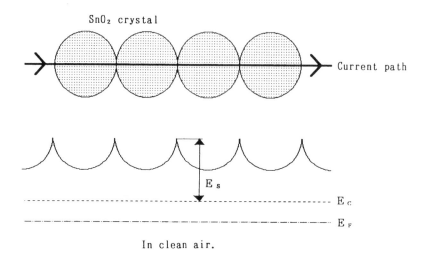

SnO₂ crystal

Current path

E s

E c

E F

In clean air.

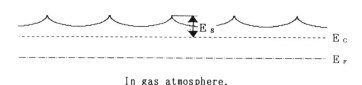

E s

E c

E F

In gas atmosphere.

FIGURE 2.22(B)   Concept of potential along current path in SnO₂ sensor (crystal diameter < twice Debye length). E_S: potential barrier; E_F: Fermi level; E_C: lowest level of conduction band; ( ): space charge layer.

the thermal excitation of charge carriers. Then, as has been explained, atmospheric oxygen adsorbs onto the high-temperature sensor surface and negative ions such as O^- are formed.[5,8] This results in the sensor resistance becoming very high. However, the dissociation of oxygen molecules needs high activation energy and involves long time constants, which account for the rise in resistance occurring after the initial rapid fall. The combination of these two effects results in the characteristic initial action transient observed.[12]

In a humid atmosphere, the condensation of water vapor on the sensor surface would also affect the initial action transient, but the major mechanisms would still be as described above.

When a sensor is de-energized, its temperature suddenly falls, eventually reaching room temperature, while adsorbed negative oxygen ions remain "frozen" on the surface. If the unenergized period is sufficiently short, a subsequent initial action transient is hardly observable upon reenergization, because most of the ions remain *in situ*. Conversely, when the unenergized period is long, the ions are desorbed or transformed to other ionic forms. Then, when the sensor is reactivated, the sequence of events leading to ionic equilibrium reappears and a longer initial action transient is observed, which explains the behavior depicted in Figure 2.24.

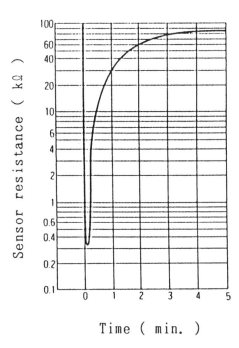

FIGURE 2.23 The short-term or "initial action" transient for a TGS 822 left unenergized for a long period (From Figaro Engineering, Inc., Catalog data.)

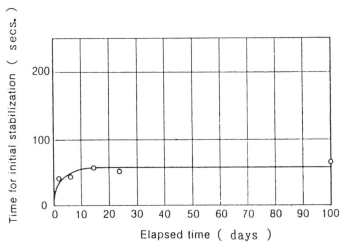

FIGURE 2.24 The relationship between initial stabilization time and storage time. Sample: TGS 109 (average of 3 samples); storage condition: in natural atmosphere and unenergized; initial stabilization time: approximate time required from switch-on for sensor resistance in air to reach 4 kΩ. (From Figaro Engineering, Inc., Catalog data.)

The same phenomena occur in combustible gases and so modify the characteristic resistances in these gases, as is shown in Figure 2.25. However, when gas is present, the initial action transient is less marked than in clean air, which suggests that the assumed sensitivity mechanism in gas — that of changes in oxygen adsorption equilibrium caused by the gas — is indeed in operation,

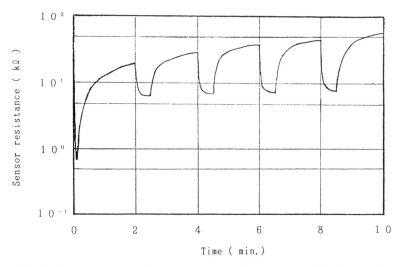

FIGURE 2.25   Response to gas during short-term transient. Sample: TGS 822; gas: ethanol vapor 300 ppm.

whereas the resistance in air is defined by the absolute density of adsorbed oxygen alone.

It should be noted here that the effects of the initial action transient are most important when such sensors are to be incorporated into battery-operated equipment where intermittent operation is expected.

2.6.2 The Long-Term Transient

When the characteristics of a stannic oxide sensor have been stabilized over a long period in clean air and it is then de-energized, its resistance normally rises, also over a long period. When the sensor is again energized, then (neglecting the initial action transient) its resistance again falls slowly to the stable value.

To illustrate this phenomenon in a graphical manner, Figure 2.26 has been prepared.[14] Here, a sensor is allowed to reach equilibrium in the energized state (region A), and is then de-energized. The subsequent rise in sensor resistance appears somewhere in region B (dashed lines), depending on how long the sensor remains in the unenergized state (from 10 to 180 days). The recovery period is the long-term transient and this is represented in region C. An equivalent set of curves measured in either very dry air or in air contaminated by an organic solvent vapor, for example, would appear in regions enclosed by the dotted lines in Figure 2.26.

In general, a longer unenergized period results in a higher resistance and an extended long-term transient. Furthermore, the long-term transient is dependent also upon the type of sensor — typically, the Figaro TGS109 requires 5 to 30 days, whereas the TGS813 needs only 3 to 10 days to achieve stability.

The long-term transient is also affected by humidity, as the following test shows,[14] in which four groups of ten sensors were taken, and all readings represent an average for the ten.

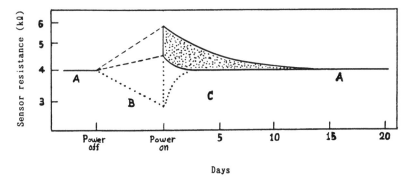

FIGURE 2.26 Conceptual figure of the long-term transient of the sensor resistance in gas. A: stabilized period for energized sensor; B: unenergized period (for 10 to 180 days, at 0 to 25°C, absolute humidity at 4 to 20 g/kg); C: long-term transient period; . . . .: behavior of sensors de-energized in contaminated atmosphere; test conditions: $V_C = 10$ V, $R_L = 4$ kΩ, $V_H = 5$ V.

Firstly, all the sensors were stabilized by long-term energization in clean air.

Secondly, three of the groups were de-energized and stored for various periods in clean air at different humidity levels, viz. (1) 25°C and 100% RH, (2) 20 ± 5°C and 50 to 65% RH, and (3) 25°C and ≥2% RH. The fourth, or control group, was stored in an energized condition in clean air as in (2) above.

Thirdly, all the sensors in the three experimental groups were energized for 15 minutes in clean air, again at 20 ± 5°C and 50 to 65% RH as in (2) above, along with the already energized sensors in the control group.

Finally, 1000 ppm of isobutane was added, and the resistances of all the sensors were measured. For the three experimental groups, and after zero days storage, the initial resistance was termed R_0, and after all other storage periods, R_S. For the control group, the initial resistance was termed R_{0-c}, and for all other (energized) storage periods, R_{S-c}. The fractional change in sensor resistance when the isobutane was added (as an average of the ten sensors per group) is plotted in Figure 2.27 against the storage time in clean air at the various humidity levels. Here, the fractional change in resistance has been normalized to that of the control group by defining it as $(R_S/R_0)/(R_{S-c}/R_{0-c})$. Thus, Figure 2.27 shows that after storage under high humidity conditions, the resistance of an unenergized sensor increases with storage time, whereas for storage under low humidity conditions, it falls; both compared with a similar sensor energized in a normally humid atmosphere at room temperature.

The mechanism underlying this phenomenon may be that while a sensor is unenergized, water molecules react with the surface to form hydroxyl groups, so that the resistance is increased. When the sensor is reenergized and its temperature rises, these surface hydroxyl groups gradually desorb and their surface concentration decreases to an equilibrium level defined by the ambient humidity and the operating temperature. Hence, the sensor resistance decreases to a value defined largely by the final hydroxyl group concentration.

The reason the time needed for stabilization is so long is because hydroxyl group formation and subsequent desorption as water molecules are processes requiring high activation energies, in contrast to the physisorption of water. Also, the reason why different types of sensors exhibit different long-term

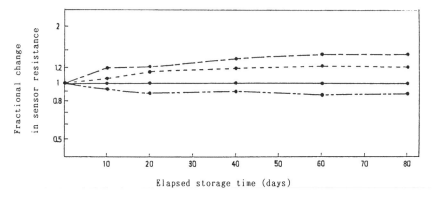

FIGURE 2.27   The influence of storage at various humidity levels on the TGS 109 sensor (averages of ten sensors per group). Test gas: 1000 ppm isobutane; (●———●): continuously energized in 20 ± 5°C, 50 to 65% RH; (●-------●): de-energized in 20 ± 5°C, 50 to 65% RH; (●———-——●): de-energized in 25°C, 100% RH; (●———-——●): de-energized in 25°C, 2% RH or less; fractional change in sensor resistance = $(R_S/R_O)/(R_{S-C}/R_{O-C})$. R_S: sensor resistance, R_O: initial sensor resistance, R_{S-C}: resistance of continuously energized sensor, R_{O-C}: initial resistance of continuously energized sensor.

transients is that they operate at different temperatures and with different ceramic surface temperature distributions.

The long-term transient is also a factor which must be carefully taken into account when the stannic oxide sensor is being considered for specific system applications.

2.7   DRIFT AND SENSITIVITY

In both storage and normal operation, the resistance of a sensor in air, and also in gas, may change over long periods and also as a result of contamination. These phenomena will be considered in Section 2.8, but prior to this, the concept of sensitivity must be examined further and in a circuit context in order to avoid certain points of confusion.

Suppose that a sensor is connected into an alarm circuit such that when its resistance in gas falls to 5 kΩ (for example), the alarm sounds. If the resistance characteristics in both air and gas are as shown in Figure 2.28(A) by the solid lines, then the alarm will operate at a gas concentration of 2000 ppm. However, should the resistance in gas R_s drift downwards, resulting in a new characteristic as shown by the lower dashed line, then the alarm would operate at 1500 ppm.

If the resistance in air R_a had drifted downwards *proportionally*, as shown by the upper dashed line, then the sensitivity as defined by the ratio R_a/R_S would have remained unchanged. Note that because a log scale of resistance has been used, a fixed vertical interval corresponds to a constant ratio at any point on the horizontal, and this is represented in the (special) case of Figure 2.28(A). The general case is that drift in R_a is not proportional to that in R_S, and this constitutes a change in sensitivity, as is illustrated in Figure 2.28(B). Here, the alarm level has again shifted from 2000 to 1500 ppm, but the sensitivity has changed in this case.

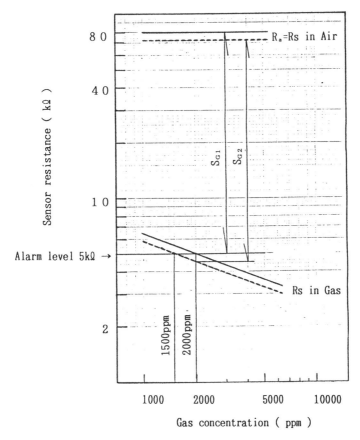

FIGURE 2.28(A) Sensor resistance drift without sensitivity change. S_{G1}: sensitivity before drift; S_{G2}: sensitivity after drift; here, $S_{G1} = S_{G2}$.

The important point to note here is that although it might be colloquially said that the sensor has become "more sensitive", this is actually in the context of the circuit as an entity, and is not necessarily true in terms of the proper definition of the phrase as applied to the sensor alone. It is for this reason that the colloquial form has been avoided in the following section, which deals with the various forms of drift in sensor resistance.

2.8 DRIFT WITH TIME

The most serious point to be addressed with respect to the lifetime of a sensor is how long that sensor will continue to function within the specifications of the system in which it is operating. Fortunately, for the properly fabricated and properly utilized stannic oxide gas sensor, this lifetime is very high — certainly some fifteen years — as is evidenced by Figure 2.29.

Conversely, catastrophic failure can occur for several reasons, including improper chemical preparation (such as contamination by chlorine or sulfur

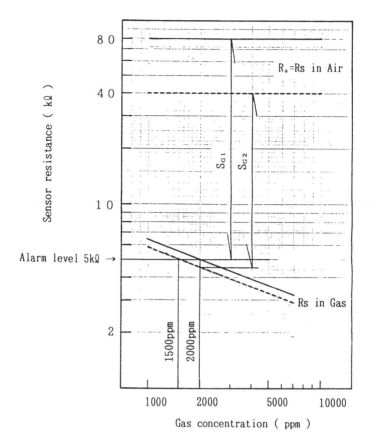

FIGURE 2.28(B)   Sensor resistance drift with sensitivity change. S_{G1}: sensitivity before drift; S_{G2}: sensitivity after drift; here, $S_{G1} \neq S_{G2}$.

dioxide, which form acids with atmospheric moisture) and mechanical shock. In both cases, damage to the lead wires or heater filament may result and has been found to be much more common than damage to the active material itself.

However, there are two phenomena which appear during the long-term observation of continuously active sensors. The first is that the resistance in gas may appear to undergo a periodic change; the second is that there is a very long-term decrease in this resistance. These changes are obviously of importance in many applications, for they can result in incorrect alarm or monitoring indications — and they may or may not be accompanied by sensitivity variations, as explained in Section 2.7.

2.8.1 Seasonal and Environmental Drift of Sensor Resistance in Gas

Some very long-term tests on early Figaro sensors have been carried out, and the results of the fifteen-year evaluation previously mentioned appear in Figure 2.29.[14] Here, two samples of early TGS 109 prototypes were continuously

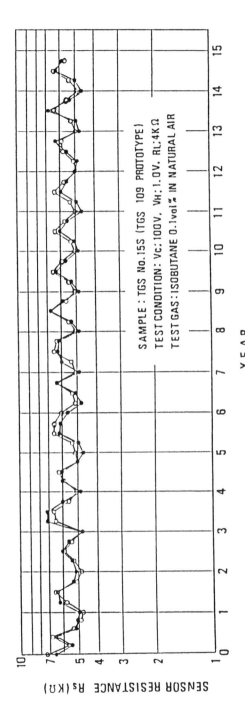

FIGURE 2.29 The long-term operation of two TGS 109 sensors showing periodic drift due to annual climate changes.

energized under normal operating conditions in the clean natural air of Japan's Kansai district, where the temperature fluctuates between a monthly average of 5.8 to 28.0°C and the humidity between monthly averages of 0.52 and 2.61% water vapor, both depending on the season. At intervals indicated by the points on the graph, the sensor resistances were measured in 1000 ppm of isobutane, and it is immediately apparent that during the warm and comparatively humid summer periods (indicated by the vertical graduations), the sensor resistances became lower than during the cool and dryer winter periods.

This pattern is also seen in Figure 2.30,[14] which relates to average sensor resistances for 100 samples of the more recent Figaro type 813. Here, the test gas was 1000 ppm of methane and the period of the test (to date) is 5 years. The expanded scale of this graph compared with Figure 2.29 again shows that the resistance in gas is lower in summer than in winter.

This seasonal drift is of considerable importance in long-term practical applications, such as permanently energized gas alarms, for the relevant instruments will alarm at slightly different gas concentrations depending upon the time of year, given a cyclic climate. Clearly, for more variable climates, weather-associated changes are equally important.

When a sensor is first installed, therefore, neither seasonal nor environmental drifts should be mistaken for any inherent long-term changes, and in the most critical applications, both the temperature and humidity should be controlled. (In this context, the temperature of the sampled or flowing ambient air is easy to control and a high humidity can be maintained by operating the sensor in close proximity to a wet wick.)

2.8.2 Long-Term Drift Due to Atmospheric Contamination

The true long-term drift characteristic considered in Section 2.6 is rarely problematical compared with either seasonal or environmental drifts or drifts resulting from contamination. This latter has been frequently encountered during long-term operation, particularly where hydrogen and alcohol vapor detection has been

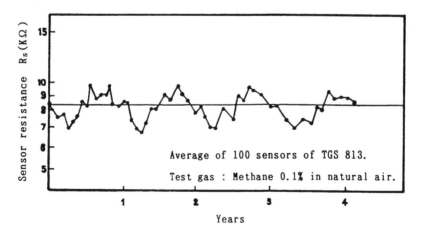

FIGURE 2.30   Long-term operation of the TGS 813 (average of 100 samples).

involved and where temperatures over 40°C and values of RH over 80% have existed. Such extremes are common in the upper corners of restaurant kitchens, for example, where gas alarms are commonly sited and which have occasionally led to serious and irreversible changes in the relevant sensors.

The results of a long-term test to illustrate contamination drift are given in Figure 2.31,[17] which relates to a sensor operating in air for 1000 days. At intervals, its resistance was measured not only in air, but also in 3500 ppm of hydrogen, then in 3500 ppm of methane. A datum resistance R_0 was taken as that at the commencement of the test in the 3500 ppm methane, and resistance ratios R_S/R_0 were plotted as shown.

These results show that the sensor resistance in air has drifted downwards approximately proportionally to that in the hydrogen, which means that the gas sensitivity to hydrogen has not changed greatly. However, had the sensor been incorporated into a hydrogen alarm circuit, it would have operated that alarm at progressively lower concentration levels.

Conversely, because the drift of resistance in methane is proportionally much less than that in air, the sensitivity has become progressively lower; yet the alarm level has varied only a little compared with that in hydrogen.

These two observations provide good examples of the terminological problems outlined in Section 2.7.

Contamination drift can be much improved by carefully defining the sintering conditions so as to inhibit crystal growth in use[16] and by optimizing the degree of polymerization of the ethyl silicate binder.[17] However, the most marked improvement is made by incorporating trace amounts of vanadium and rhenium into the sensor,[18] and this is clearly seen in Figure 2.32.

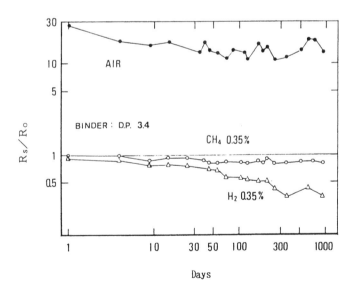

FIGURE 2.31 Drift of sensor resistance in H_2 for long-term operation. R_0: sensor resistance in 3500 ppm CH_4 at commencement of test. (From Yasunaga, S., Proc. Transducers '85, Philadelphia, 1985, 393.)

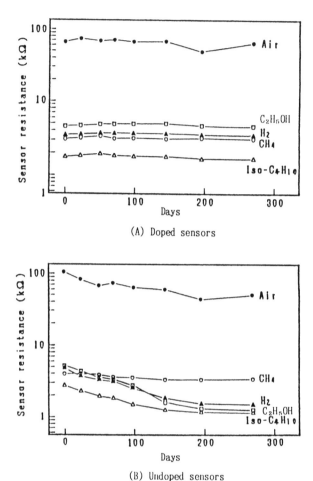

(A) Doped sensors

(B) Undoped sensors

FIGURE 2.32   Long-term stability of 1.46 mol% Re/0.71 mol% V doped sensors (A) and undoped sensors (B). Gas conc.: 3500 ppm; sensor temp.: 350°C. (From Matsuura, Y, et al., *Sensors Actuators,* 44, 223, 1988. With permission.)

Here, the actual values of R_a and R_s in several different gases are recorded for sensors both undoped (diagram A) and doped with rhenium and vanadium (diagram B). Again, the energized sensors were maintained in air and extracted at intervals for resistance measurement in 3500 ppm of several gases.

Diagram A shows that over a period of 300 days, the doped sensor exhibited almost no contamination drift as compared with the undoped sensor of diagram B.

2.8.3 Contamination Drift in Practice

Figure 2.33 is a graphic depiction of the effects of contamination drift and has been compiled in the light of a number of actual cases. The solid line shows that the resistance of a sensor R_s can fall with respect to its initial value R_0 during operation under adverse conditions, but also how it may recover when subsequently stored unenergized in clean air. The dashed line, however, illustrates that if operated for a long period in very adverse conditions, the subsequent recovery is only partial, that is, some irreversible damage has occurred.

In this context, "adverse" and "very adverse" conditions imply extremes of ambient temperature and humidity coupled with continuous atmospheric pollution. A case in point would be where a sensor is installed in a town gas alarm instrument and is calibrated to alarm at a given concentration of hydrogen, which is a constituent of town gas. Following the earlier example where the instrument might be installed in an upper corner of a commercial kitchen, it is here where any unburnt hydrogen would be expected to collect. However, it is also here where hot air and cooking vapors also collect, and where the atmosphere might reach extremes of temperature and humidity, such as 40°C and over 80% RH, as previously mentioned.

The directly heated form of sensor is particularly vulnerable under such conditions, for although its normal operating temperature might be about 350°C, this might rise to some 450 to 550°C under the continuous action of the reducing gases in the pollutant vapors, and of any hydrogen which may be present below the alarm level. It is such a combination of high temperature, humidity, and contamination which is thought to be responsible for most of the observed changes in resistance — and hence alarm points — for sensors working in moderate to highly adverse conditions.

2.8.4 Mechanisms of Contamination Drift

Considerable work has been carried out on this topic, and though not yet fully understood, some phenomena believed to occur have been elucidated:

1. The hydrogen oxidation activity (or conversion rate) falls;[17,18] Figure 2.34[19] shows this for sensors with and without binder, and with and without the addition of palladium, a common additive the effects of

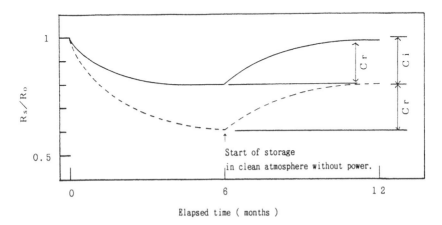

FIGURE 2.33 Drift patterns of sensor resistance in H_2 under adverse operating conditions. R_s: sensor resistance; R_o: initial value of sensor resistance; ———: operation under adverse conditions; ------: operation under very adverse conditions; C_i: irreversible change under very adverse conditions; C_r: reversible change under very adverse conditions.

which will be treated in the next chapter. Concomitantly, the gas sensitivity rises with respect to its initial value.

2. The amount of water (as a surface hydroxyl group) which desorbs in the high temperature region of the TPD spectrum decreases. Figure 2.35[19] is such a spectrum, and it shows physisorbed water desorbing at about 100°C followed by the hydroxyl group desorbing above 400°C; both decrease after long-term energization, particularly at higher operating temperatures.

3. The amount of oxygen (as O^- or O^{2-} adsorbed ion species) which desorbs over the temperature range 420 to 600°C of the TPD spectrum decreases after long-term energization, as is also seen in Figure 2.35.[19] Above this range, it is the lattice oxygen which desorbs.

4. Figures 1.24 and 1.27 have indicated that sensor resistance and sensitivity are related to crystal size and the extent of sintering. However, though it may be thought that continued crystal growth or sintering during operation might lead to long-term drift, this has been discounted by Iwamoto[8] and Matsuura,[19] who report that any such changes are too small to account for long-term sensitivity drift.

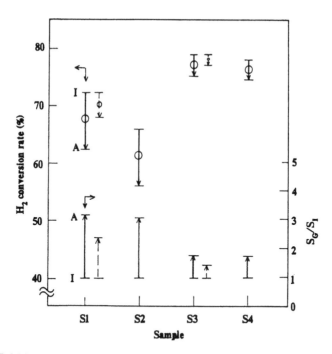

FIGURE 2.34 Changes in the catalytic activity and sensitivity of sensors over long-term operation. (Note: a high conversion rate implies a high catalytic activity). I: initial time; A: after 6 months; S_I: initial sensitivity; S_G: sensitivity after 6 months; (-⊖-): conversion rate change for sensor at 450°C; (——): sensitivity change for sensor at 450°C; (-⊖-): conversion rate change for sensor at 350°C; (-----): sensitivity change for sensor at 350°C. Samples: S1: sensor with binder and Pd; S2: sensor with binder; S3: sensor with Pd; S4: Sensor without binder or Pd. (From Matsuura, Y. et al., *Sensors Actuators, 44*, 223, 1988.)

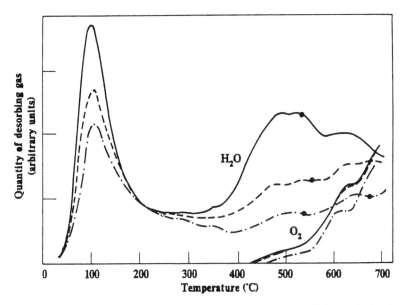

FIGURE 2.35 Temperature-programmed desorption (TPD) spectra of H_2O and O_2 desorbed from sensors with binder after pretreatment as follows: (1) heating to 550°C in helium for 30 min; (2) heating to 550°C in oxygen/helium mixture for 30 min; (3) heating to 400°C in humid 10%-oxygen-in-helium mixture for 60 min; and (4) cooling to room temperature. (———): Desorption spectrum of O_2 for sensor (1) which was unenergized prior to pretreatment; (—•—): ditto for H_2O; (-----): desorption spectrum of O_2 for sensor (2) after energization at 350°C for 6 months; (--•--): ditto for H_2O; (•—•—•): desorption spectrum of O_2 for sensor (2) after energization at 450°C for 6 months, and (•—•—•): ditto for H_2O. (From Matsuura, Y. et al., *Sensors Actuators*, 44, 223, 1988.)

Observations (1), (2), and (3) lead to some possible explanations of sensor resistance decrease with time, as follows.

When the hydrogen oxidation activity is high, oxidation takes place wholly at the surface of the ceramic, and the process does not penetrate the body of the sensor. Hence, the surface resistance changes, but the bulk resistance does not, so that the overall resistance change is not great. However, when the oxidation activity is low, some hydrogen can penetrate the ceramic, leading to a change in bulk resistance, too. Hence, the overall resistance change eventually becomes large.

The fall in the catalytic activity of the ceramic has been investigated by Yasunaga et al,[17] who consider the hydrogen oxidation activity of amorphous silica formed on the ceramic surface by the binder. If this becomes hydrated, its activity is approximately quadrupled, and it is noteworthy that sensors become hydrated during unenergized storage and also exhibit high activity immediately after energization, which falls as dehydration takes place at working temperatures.

Matsuura et al.[19] have suggested that the density of the surface hydroxyl groups affects the oxidation rate of hydrogen because the hydroxyl groups already existing on the surface obstruct the formation of new hydroxyl groups which would be intermediate products in this oxidation process. Hence, a fall in catalytic activity should be brought about by the decrease in hydroxyl group density at working temperatures.

These hypotheses are acceptable in the light of both reversible and nonreversible resistance drifts which occur under adverse working conditions. However, if the decrease in oxygen desorption noted in (3) above results not only from a fall in surface hydroxyl group density, but also in surface defect density (which is where oxygen is adsorbed), this should also contribute to irreversible drift. This has not been fully investigated at the time of writing, however. Iwamota[8] has suggested that the action of rhenium and vanadium additives affect the oxygen adsorption sites, but has not considered the phenomena of item (3) above. It is likely that a full investigation of the action of these additives could elucidate the rather obscure mechanisms of resistance drift considerably.

REFERENCES

1. Kubokawa, Y., et al., The electrical conductivity change caused by the chemisorption of hydrogen on ZnO, ZnO$\cdot$Cr$_2$O$_3$ and ZnO$\cdot$MoO$_3$, *J. Phys. Chem.*, 60, 833, 1965.
2. Tarama, K., On the catalysis of the transition metal oxides, *Yuuki-Gousei-Kagaku (Organic Synthesis Chemistry)*, 16, 433, 1958.
3. Seiyama, T., et al., A new detector for gaseous components using semiconductive thin film, *Anal. Chem.*, 34, 1502, 1962.
4. Seiyama, T., et al., *Kagaku Sensors (Chemical Sensors)*, chap. 2, Koudansya, 1982.
5. Yamazoe, N., et al., Interactions of the tin oxide surface with O$_2$, H$_2$O and H$_2$, *Surf. Sci.*, 86, 335, 1979.
6. Ihokura, K., Effects of tetraethyl silicate binder on a SnO$_2$ gas sensor, *Denki Kagaku (J. Electrochem. Soc. Jpn.)*, 51, 773, 1983.
7. Mizokawa, Y., et al., ESR study of adsorbed oxygen on tin dioxide, *Ouyou Buturi (Jpn. J. Appl. Phys.)*, 46, 580, 1977.
8. Iwamoto, M., Adsorption and desorption of gases on metal oxides and their sensor functions, Proc. 34th Meet. Jpn. Assoc. Chem. Sensors, Electrochem. Soc. Jpn., August 1989, 23.
9. Egashira, M., An overview of semiconductor gas sensors, Proc. Symp. Chem. Sensors, 1987 Joint Congress, Electrochem. Soc. Jpn. and U.S., Honolulu, October 1987, 39.
10. Seiyama, T., et al., Gas detection by activated semiconductive sensors, *Denki Kagaku* (J. Electrochem. Soc. Japan), 40, 244, 1972.
11. Seiyama, T., Kinzoku-sankabutu to sono syokubai-sayou (Metal oxides and their catalytic actions), Koudansya, 1978, 179.
12. Ihokura, K., A tin dioxide gas sensor for deoxidizing gases, *New Mater. New Process. Electrochem. Technol.*, 1, 43, 1981.
13. Ri, S., et al., Debye length and gas sensor mechanism of undoped and Sm$_2$O$_3$-doped ZnO ceramics, *J. Ceramic Soc. Jpn.*, 94, 419, 1986.

14. Figaro Engineering Inc., Technical information: Basis of Tin Dioxide Gas Sensor.

15. Ihokura, K., Research on sintered tin dioxide gas sensors for combustible gases, Ph.D. thesis, Kyushu University, 1983.

16. Nakamura, Y., et. al., Stabilization of SnO_2 Gas Sensor Sensitivity, Proc. 2nd Int. Meet. Chem. Sensors, Bordeaux, 1986, 163.

17. Yasunaga, S., et al., Effects of tetraethyl orthosilicate binder on the characteristics of SnO_2 ceramic type semiconductor gas sensor, Proc. Transducers '85, Philadelphia, 1985, 393.

18. Matsuura, Y., et al., Stabilization of the SnO_2 Gas Sensor, Tech. Dig. 9th Sensor Symp., Tokyo, 1990, 91.

19. Matsuura, Y., et al., Mechanism of gas sensitivity change with time of SnO_2 gas sensor, *Sensors Actuators,* 44, 223, 1988.

20. Figaro Engineering, Inc., Catalog data.

CHAPTER THREE

Sensitivity Modification Using Additives

3.1 THE NEED FOR ADDITIVES

A gas sensor composed only of stannic oxide has a limited sensitivity to chemically stable gases such as methane, which is unfortunate, particularly in the context of inexpensive instrumentation. Furthermore, again in terms of practical application, a higher degree of sensitivity to certain gases — that is, selectivity — may be desirable.

For a wide spectrum instrument (such as would be required for the detection of general fuel gases), similar sensitivities to several relevant gases would be advantageous. However, Figures 2.2 and 2.5 indicate that the stannic oxide sensor is less sensitive to both methane and isobutane than it is to hydrogen, and furthermore, the sensitivity is actually highest for ethanol, which is often considered to be a "noise", "nuisance" or "interference" gas.

Because the gas sensitivity is closely related to redox reactions of the detected gases on the sensor surface, it is reasonable to suppose that it could be improved by including additives which act as catalysts to these reactions. According to Taguchi[1] and Shaver,[2] both palladium and some other inclusions have this effect, and indeed commercial sensors do usually contain additives such as these for sensitivity enhancement. This technique will now be considered.

3.1.1 The Addition of Palladium

A convenient procedure for adding palladium is to first dissolve the metal in aqua regia and then dilute with distilled water to define a prescribed palladium concentration. This solution is then added to stannic oxide powder (prepared as described earlier) prior to its calcination. It is also convenient to add the same volume of palladium solution to a fixed quantity of stannic oxide powder each time and to control the amount of palladium added via the degree of dilution. The resulting mix must be thoroughly stirred prior to drying and final calcination.

3.1.2 The Effect of Palladium on Sensitivity

Sensor temperature characteristics for stannic oxide ceramic containing varying amounts of palladium from 0 to 2.0% are shown in Figures 3.1 through 3.5.

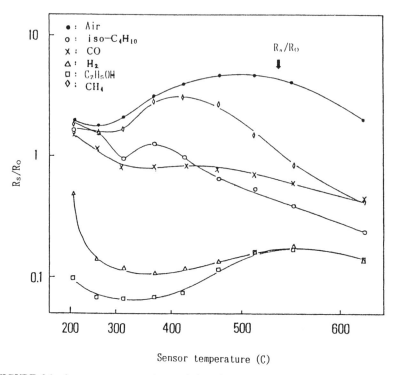

Sensor temperature (C)

FIGURE 3.1   Sensor temperature characteristics of a stannic oxide sensor in 1000 ppm of various gases (diagram 1 — no added palladium). R_o: sensor resistance at 450°C in 1000 ppm isobutane (1.13 kΩ); R_s: sensor resistance for each gas at various sensor temperatures.

In each case, the sensor resistance is normalized to its value at 450°C in 1000 ppm of isobutane.

An initial comparison of these figures shows that the addition of palladium has a marked effect on the sensor resistance, particularly in air and in carbon monoxide. Recalling that sensitivity is defined as the resistance in air R_a divided by that in gas R_S, it is apparent that this parameter will be affected in a manner dependent upon the fractional changes in each.

If the sensitivity is derived from Figures 3.1 through 3.5 for a temperature of 450°C in all cases, the resultant plot appears as in Figure 3.6. Here, the effect of palladium is seen to be highly dependent upon the gas involved, and the sensitivities to methane, isobutane and hydrogen, for example, are all enhanced, with a fairly well-defined optimum percentage of palladium for each. In the case of carbon monoxide, this optimum is about 0.2% with a secondary, smaller maximum at about 1.0%, which is quite a complex effect. For ethanol vapor, the sensitivity decreases when palladium is added, though it is rather better at 1.0% than at 0.5%, again a complex response.

The general conclusion is that for sensors intended for the detection of fuel gases such as methane, the addition of about 1.0% of palladium is a valid technique for improving sensitivity. However, some specific effects of palladium may be also be derived from Figures 3.1 through 3.6 and listed as follows:

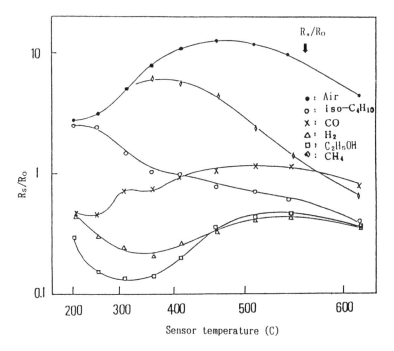

FIGURE 3.2 Sensor temperature characteristics of a stannic oxide sensor in 1000 ppm of various gases (diagram 2 — 0.2% added palladium). R_o: sensor resistance at 450°C in 1000 ppm isobutane (0.48 kΩ); R_s: sensor resistance for each gas at various sensor temperatures.

1. At lower temperatures (ca. 200°C), the addition of palladium affects only the sensitivity to carbon monoxide (for the group of gases shown).
2. For carbon monoxide, the greatest effect is provided by 0.2% palladium addition at these low temperatures (around 200°C).
3. No effect is seen for methane below a working temperature of about 300°C.
4. The sensitivities to hydrogen and ethanol vapor are initially high, particularly for low temperatures, and the addition of palladium effects no improvement.
5. The most marked change is the resistance in air, for which the local maximum peak not only becomes sharper and higher, but shifts towards lower temperatures as the percentage of palladium is increased. The greatest value of this local maximum appears for the addition of 1.0% of palladium.
6. For an operating temperature of 450°C, Figure 3.6 shows that for all the gases, the sensitivity falls as the palladium content rises above about 1.0%.

3.1.3 The Effects of Palladium on Sensor Response Speed and Short-Term Transient

1. The addition of palladium is also useful in improving sensor response speed, as is shown by Figure 3.7, which compares sensors without

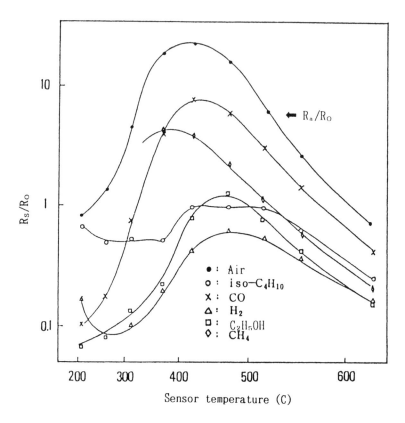

FIGURE 3.3   Sensor temperature characteristics of a stannic oxide sensor in 1000 ppm of various gases (diagram 3 — 0.5% added palladium). R_o: sensor resistance at 450°C in 1000 ppm isobutane (4.28 kΩ); R_s: sensor resistance for each gas at various sensor temperatures.

palladium and with 1% added palladium, both working at 450°C. Each is suddenly immersed in 1000 ppm of isobutane, then withdrawn after 4 minutes. Both transients are seen to be faster in the case of the sensor with added palladium, particularly in terms of the recovery time. (This effect is continuously enhanced with up to at least 2% of added palladium.)

2.   The short-term transient has been described in the previous chapter, the initial stabilization time being the time needed for a sensor to achieve stability after a storage period. This can be markedly shortened by the addition of palladium, as is shown in Figure 3.8, where again a sensor with 1% of palladium is compared with a sensor without palladium. (This effect increases until at least 2% of palladium is reached).

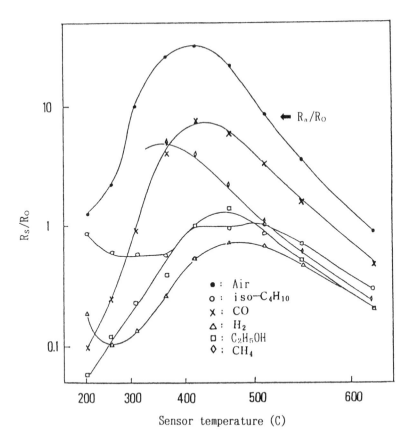

FIGURE 3.4 Sensor temperature characteristics of a stannic oxide sensor in 1000 ppm of various gases (diagram 4 — 1.0% added palladium). R_o: sensor resistance at 450°C in 1000 ppm isobutane (9.65 kΩ); R_s: sensor resistance for each gas at various sensor temperatures.

3.1.4 The Effects of Palladium on Gas Concentration Characteristics and Sensitivity

Figure 3.9 gives a family of normalized gas concentration characteristics for a sensor without added palladium, and working at 450°C, for comparison with that for a sensor with 1.0% of added palladium working under the same conditions shown in Figure 3.10. These diagrams may also be used to compare gas sensitivities because (as was explained in Section 1.5), in graphs with logarithmic ordinates, the difference between the air level R_a/R_0 and any resistance ratio R_s/R_0 is a measure of the relevant gas sensitivity.

Several conclusions may be drawn from such comparisons. When palladium is added:

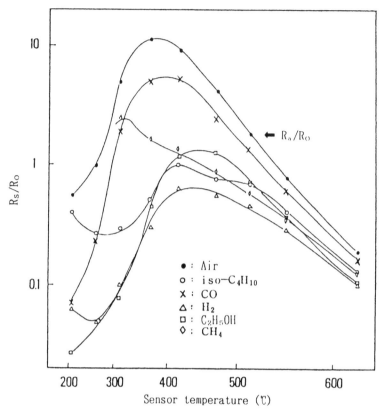

FIGURE 3.5 Sensor temperature characteristics of a stannic oxide sensor in 1000 ppm of various gases (diagram 5 — 2.0% added palladium). R_o: sensor resistance at 450°C in 1000 ppm isobutane (18.6 kΩ); R_s: sensor resistance for each gas at various sensor temperatures.

1. The sensitivities to methane and isobutane are significantly improved.
2. The sensitivity to methane becomes higher than that to carbon monoxide.
3. The sensitivity to ethanol vapor becomes somewhat lower that those to isobutane and methane.
4. The sensitivities to methane and isobutane and hydrogen move closer to those for hydrogen and ethanol vapor.

For a commercial fuel gas sensor, item 4 is usually a desirable effect because gas alarm instruments are basically used for determining when small fractions of the lower explosive limit (LEL) concentration have been reached for any gas or mixture of gases. However, in this same context, relative insensitivity to "noise" gases is also desirable, and this is also provided by palladium addition according to item 3. It is for reasons such as these that the addition of palladium provides one effective method for the realization of commercial fuel gas sensors.

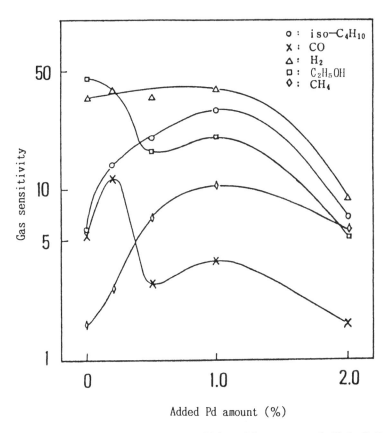

Added Pd amount (%)

FIGURE 3.6 The relationship between gas sensitivity and the percentage of added palladium. Gas sensitivity: R_a/R_s; gas concentration: 1000 ppm; sensor temperature: 450°C.

3.1.5 The Influence of Water Vapor on Sensors with Added Palladium

Work described in Section 2.3.2 led to Figure 2.11, which is repeated here as Figure 3.11. This is a family of normalized resistance characteristics for a sensor in the presence of both gas and water vapor, and is included for comparison with Figure 3.12, which is relevant to a sensor with 1.0% added palladium. Both sensors worked at 450°C in an atmosphere at 20°C, and the concentration of each gas was maintained at 2000 ppm. In both cases, a curve labelled "H_2O" represents the case where only water vapor, but no gas, was present.

The comparison shows that for air with water vapor alone (humid air) the sensitivity of the sensor with 1.0% palladium falls by about half that for the sensor without palladium. This sensitivity S_w is shown in both figures (for 20,000 ppm H_2O) so that the reduction can be clearly seen.

However, if a gas is also present, the overall sensitivity is not simply the sum of the sensitivity to water vapor plus that to the gas. This is unfortunate

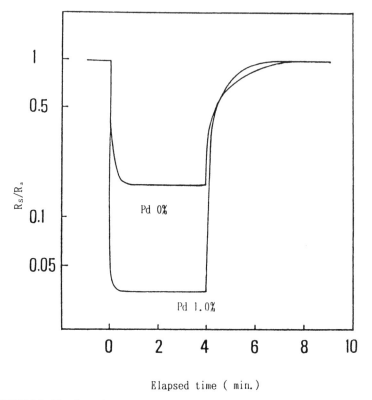

FIGURE 3.7   The change in response speed caused by palladium addition. Gas: isobutane 1000 ppm; R_s: sensor resistance; R_a: sensor resistance in air; sensor temperature: 450°C.

because the addition of palladium would be expected to lower the humidity dependency in the presence of gas. In fact, the phenomena which take place are complex and do not necessarily lead to such a reduction, a case in point being that for carbon monoxide.

The relative sensitivity to carbon monoxide in humid air compared with that in dry air is shown as HD (where HD means "Humidity Dependency") in both figures. The value of HD is much larger for the sensor with palladium than for the sensor without palladium even though the change in sensitivity to water vapor alone is the converse of this, the value of S_w being smaller for the sensor with palladium than for that without. If the sensitivity to carbon monoxide were not affected by humidity at all, the relevant plot would show a straight line parallel to the X-axis. Then, if the humidity dependency with carbon monoxide were the simple sum of water vapor and carbon monoxide sensitivities, the carbon monoxide plot would be parallel to the curve for water vapor, whereas the actual curves shown in the figures are quite different from this.

For a sensor without palladium, Figure 3.11 shows that the relative sensitivity to carbon monoxide in the presence of water vapor S_{CH} *decreases* from the dry air value S_{CD}. However, for a sensor with palladium, Figure 3.12 shows that this relative sensitivity *increases* from the dry air value S_{CD}.

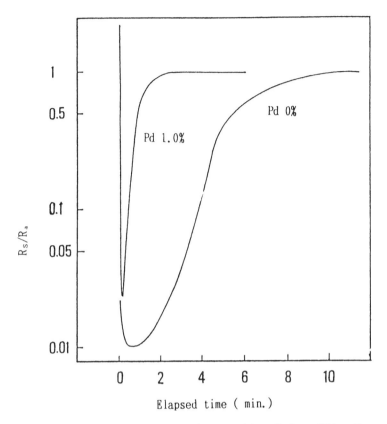

FIGURE 3.8 The change in short-term transient caused by palladium addition. R_s: sensor resistance; R_a: sensor resistance in air; sensor temperature: 450°C.

Thus, the effect of coexisting water vapor on the carbon monoxide sensitivity is much too large to ascribe to the simple addition of the two sensitivities, and a possible explanation may be rooted in an oxidation process on the sensor surface:

$$CO + H_2O \rightarrow CO_2 + H_2 \qquad (3.1)$$

when the resulting hydrogen would be sensed, too. This is the familiar water-gas reaction,[3] and analogous processes might also be at work for other gases.

The humidity-dependency of the sensitivity to hydrogen and isobutane have also increased slightly for the sensor with palladium, but that for methane has become very small.

3.2   BASIC PROCESSES IN PALLADIUM-DOPED SENSORS

The mechanisms by which the inclusion of palladium increases the gas sensitivity have not yet been elucidated, though they are thought to relate to both

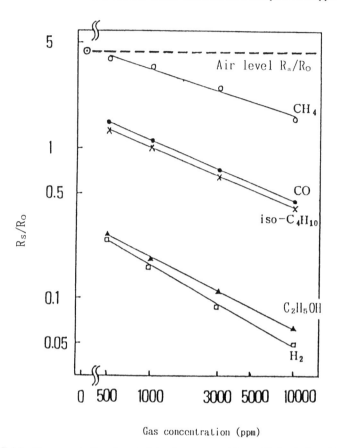

Gas concentration (ppm)

FIGURE 3.9   Gas concentration characteristics for a sensor without added palladium. R_s: sensor resistance; R_o: sensor resistance in 1000 ppm isobutane; sensor temperature: 450°C.

normal catalytic activity and also the strong metal surface interaction (SMSI) effect between the palladium and the stannic oxide surface, which may operate both individually and concurrently. Experimental support for such mechanisms is detailed below.

3.2.1 Palladium as an Oxygen Dissociation Adsorption Catalyst[4]

Figures 3.1 to 3.5 indicated that palladium doping increases the resistance of stannic oxide sensors in air. Figure 3.13 now gives actual values of sensor resistance in air R_a with various percentages of palladium, also as functions of temperature.

It can now be seen that not only does R_a rise as the palladium inclusion increases, but the peak at a characteristic temperature becomes sharper and also shifts towards lower temperatures, at least up to 1% palladium by weight. Doping levels increasing to 2% continue to shift the peaks to the left, but the absolute resistance falls somewhat.

The maximum increase represents a resistance multiplication of about 22, and the maximum peak shift is a fall of about 100°C.

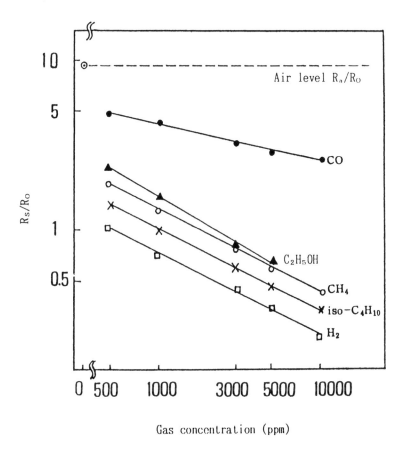

FIGURE 3.10   Gas concentration characteristics for a sensor with 1.0% of added palladium. R_s: sensor resistance; R_0: sensor resistance in 1000 ppm isobutane; sensor temperature: 450°C.

Figure 3.14 gives a family of sensor resistance curves in 2000 ppm of methane, again for various levels of palladium doping, and as functions of temperature. Here, the sensor resistance R_S is seen to rise over the whole temperature range except at very low doping levels, as evidenced by the 0.2% curve; the maximum resistance multiplication is about 7.

A comparison of Figures 3.13 and 3.14 will indicate how palladium doping improves the sensitivity, particularly at lower temperatures. The increase in R_a (Figure 3.13) is larger than in R_S (Figure 3.14), so that the sensitivity, R_a/R_S, rises concomitantly. Furthermore, this effect is most pronounced at lower temperatures as the percentage of palladium increases, because of the leftwards shift in the R_a maxima in Figure 3.13.

In Chapter 2, it was shown that R_a is dependent upon adsorbed oxygen ions, and it is in this context that the results of Figure 3.13 can be understood, because palladium catalyzes the dissociative adsorption of oxygen. In particular, it catalyses the formation of O^- and O_2^- over lower and narrower temperature ranges, so that it is not surprising that the peaks in Figure 3.13 represent maximal resistance rises at decreased temperatures.

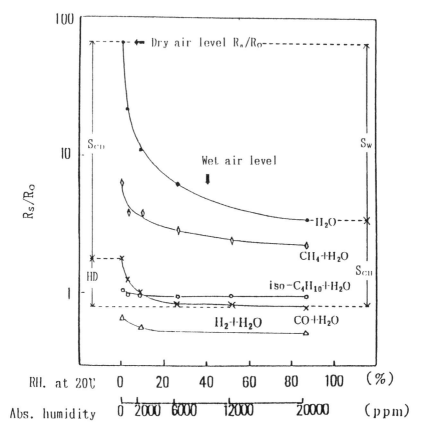

FIGURE 3.11   Sensitivity to water vapor and humidity dependency of a sensor without added palladium. Sensor temperature: 450°C; atmospheric temperature: 20°C; gas concentration: 2000 ppm in air and at various humidity levels; R_s: sensor resistance; R_o: R_s in 2000 ppm isobutane and 12,000 ppm water vapor (1.26 kΩ); S_w: sensitivity to water vapor; S_{CD}: sensitivity to carbon monoxide in dry air; S_{CH}: relative sensitivity between water vapor and carbon monoxide, (the value at 20,000 ppm water vapor is shown in the figure); HD: humidity dependency of CO sensitivity; basically corresponds to relative sensitivity from dry to humid CO (0 to 20,000 ppm water vapor). (The value at 12,000 ppm water vapor is shown in the figure.)

A mechanism proposed by Bond et al.[5] involves adsorbed oxygen spillover from palladium on the surface of the stannic oxide to the surface of that stannic oxide itself. This hypothesis is further supported by the fact that the same effects are seen when the palladium is not added in solution form, but only by the mechanical mixing of palladium black, of average grain size 40 to 50 μm.

A final item of support for the catalysis mechanisms proposed is that the initial stabilization time is shortened, as has been shown in Section 3.1.3 relating to Figure 3.7, for this is also a function of the formation of adsorbed oxygen ions.

Summarizing, it may be said that palladium dopant catalyzes the dissociative adsorption of oxygen, so increasing the negative charge density on the stannic oxide surface and, hence, increasing its resistance in air R_a, particularly at lower temperatures. Thus, the sensor becomes more sensitive to combustible gases at these lower temperatures.

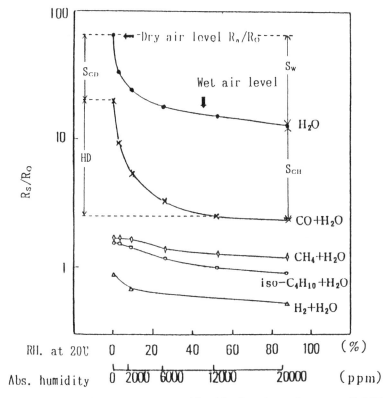

FIGURE 3.12 Sensitivity to water vapor and humidity dependency of a sensor with 1.0% of added palladium. Sensor temperature: 450°C; atmospheric temperature: 20°C; gas concentration: 2000 ppm in air and at various humidity levels; R_s: sensor resistance; R_o: R_s in 2000 ppm isobutane and 12,000 ppm water vapor (2.65 kΩ); S_w: sensitivity to water vapor; S_{CD}: sensitivity to carbon monoxide in dry air; S_{CH}: relative sensitivity between water vapor and carbon monoxide, (the value at 20,000 ppm water vapor is shown in the figure); HD: humidity dependency of CO sensitivity; basically corresponds to relative sensitivity from dry to humid CO (0 to 20,000 ppm water vapor). (The value at 12,000 ppm water vapor is shown in the figure.)

3.2.2 Palladium as an Electron Donor or Acceptor

Knowing that the work function of stannic oxide is modified by the presence of the palladium, Yamazoe et al.[6,8] postulate the following mechanisms by which the sensitivity is increased.

At normal sensor temperatures, the palladium exists in the form of an oxide, PdO, on the surface of the stannic oxide and can receive conduction electrons from that surface, so increasing its work function. This raises the grain boundary potential of the ceramic, so increasing its resistance.

When a reducing gas appears, the PdO is reduced to metallic palladium which can no longer bind the received electrons, so that the resistance decreases again. Yamazoe et al.[6] dubbed this process the "electronic effect" and summarized the mechanism as shown in Figure 3.15. In terms of the overt improvement in sensitivity to certain gases, a dopant such as palladium is sometimes called a "promoter".

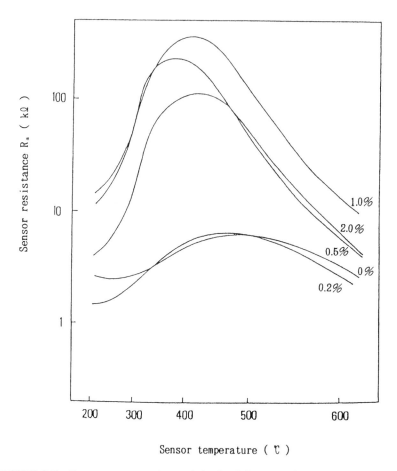

Sensor temperature (℃)

FIGURE 3.13   Sensor temperature characteristics for SnO$_2$ sensors in clean air with various palladium inclusions. (Percentages in the figure are the amounts of doped palladium by weight.)

3.2.3 The Condition and Amount of Palladium Dopant

It will now be clear that for a given gas, there will be an optimum percentage of palladium dopant in a sensor ceramic, and a plot of gas sensitivity vs. percentage dopant is given for hydrogen in Figure 3.16.[6]

Matsushima et al.[7] examined the state of the palladium on the ceramic surface using TEM and found the average particle size to be 5 nm or less, up to 3% by weight, after which particle sizes up to 10 nm were observed at 5% by weight. This finding is depicted in Figure 3.17.

At low percentages, the palladium is distributed uniformly on the ceramic surface, and its surface density increases, also uniformly, up to about 3% by weight. Hence, the abovementioned effects of palladium also rise fairly uniformly with the palladium percentage. However, when about 5% by weight of palladium is reached, cohesion between the particles occurs, and the effective

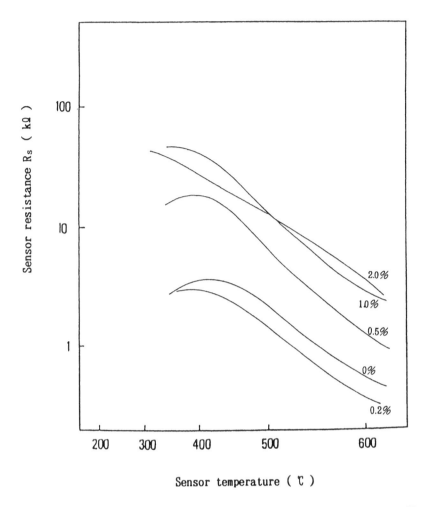

FIGURE 3.14 Sensor temperature characteristics for SnO_2 sensors in 2000 ppm methane with various palladium inclusions. (Percentages in the figure are the amounts of doped palladium by weight.)

density in terms of particles per unit area consequently decreases. Hence, the observed effects decrease again.

3.3 OTHER ADDITIVES

The effects of numerous additives have been investigated, most of these being either noble metals or various oxides; some of these are tabulated in the Introduction (Table I.1). Discussions on some of the more important additives now follow.

Type	Chemical	Electronic
Model	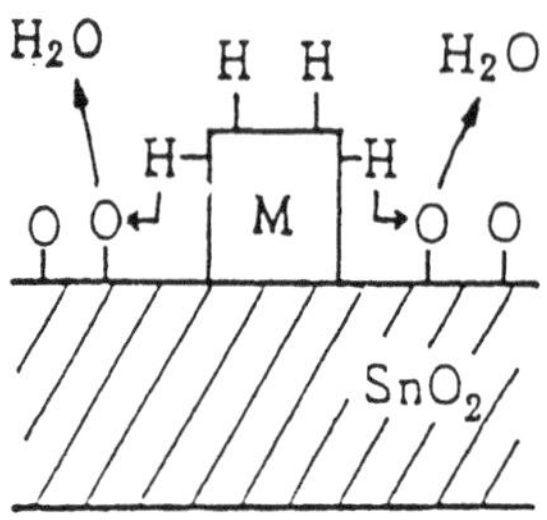 H_2O H H H_2O H—‖—H O O⤶ ‖ M ‖ ↳O O SnO_2	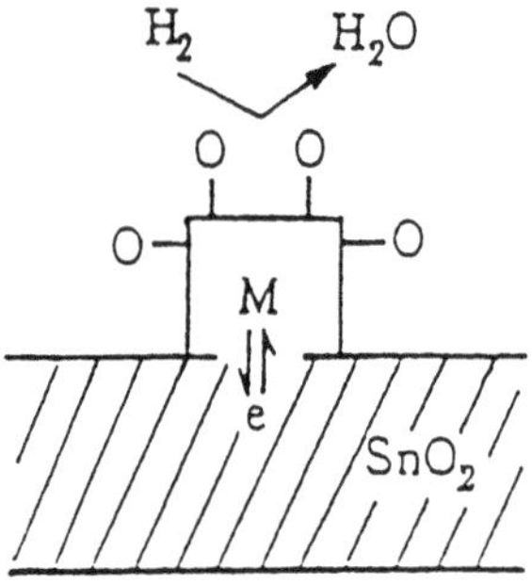 H_2 H_2O O O O—‖ M ‖—O e SnO_2
Rôle of noble metal	Activation and spill-over of sample gas	Electron donor or acceptor
Origin of conductivity change	Change of oxidation state of SnO_2	Change of oxidation state of noble metal
Example	$Pt\text{-}SnO_2$	$Ag\text{-}SnO_2$, $Pd\text{-}SnO_2$

FIGURE 3.15   Promoting effect of a noble metal in the stannic oxide sensor. (From Yamazoe, N., Proc. 3rd Int. Meet. Chem. Sensors, Cleveland, 1990, 3. With permission.)

3.3.1 Noble Metal Additives

Noble metals are generally oxidation catalysts, and the action of one of them — palladium — has been discussed.

The effects of silver and platinum have been included in Figure 3.15; the results of adding gold, rhodium, ruthenium and indium have also been reported. The influences of the various metals on the gas sensitivity are different,[8] and as examples, Figure 3.18 has been included. Here the differences between the effects of palladium, platinum and silver dopants on the sensitivity to various gases are clearly seen.

Usually, the commercial stannic oxide sensor is doped with more than one noble metal to combine their effects, making possible the design of sensors for various different applications.

3.3.2 Further Additive Techniques

Early in sensor development, the addition of antimony was used to modify sensitivity by changing the stannic oxide work function via valency control.[9]

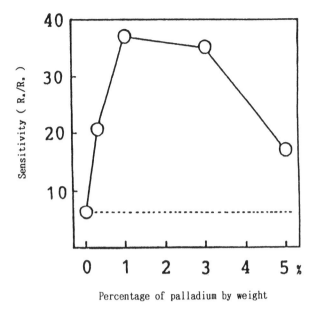

Percentage of palladium by weight

FIGURE 3.16   Gas sensitivity of Pd-doped SnO_2 sensor to 194 ppm H_2. Sensor temperature: 100°C. (From Yamazoe, N., Proc. 3rd Int. Meet. Chem. Sensors, Cleveland, 1990, 3. With permission.)

Also, sensors with many different metal oxide additives were investigated in the hope of producing complex oxides with the stannic oxide, to result in controllable sensitivity and selectivity, but none has so far proved commercially viable.

The use of alumina has already been described, and it will be recalled that it improves the ceramic strength and also the sensitivity, essentially by porosity control.[10] However, these effects are basically structurally defined and so should not be confused with the effects considered in this section.

Yamazoe[6] has proposed that the effects of various additives should be investigated in terms of receptor identification for different gases, with a view to selectivity enhancement. Maekawa and Yamazoe et al[11] have reported enhanced sensitivity to hydrogen sulfide by the addition of copper oxide; Matsushima and Yamazoe et al.[12] have reported enhanced sensitivity to alcohol by the addition of lanthanum oxide. These techniques will be considered further in the next chapter.

Finally, it will have been noted that all the methods so far described have depended upon *adsorption*. Morrison[13] has suggested that *absorption* in the bulk of an active material such as bismuth molybdate (which has a very high diffusion constant for oxygen vacancies) could provide an alternative mechanism by allowing combustible gas to extract lattice oxygens. He also notes that this might resolve problems of reproducibility, which are said to be even greater with thin or thick film techniques than with the bulk methods detailed herein.

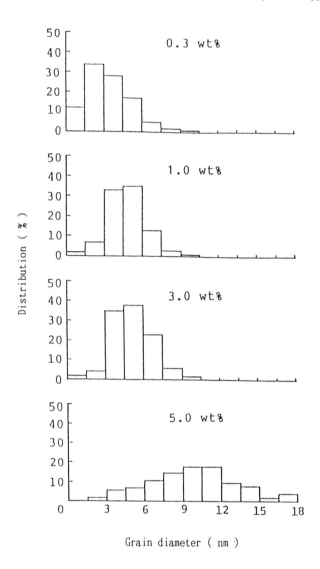

FIGURE 3.17   Grain diameter distributions for various doping amounts of palladium on SnO_2 sensor. (From Matsushima, S. et al., *Chem. Lett.,* 1989, 845, 1989. With permission.)

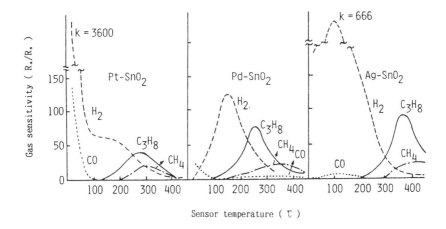

FIGURE 3.18 Sensitivities of SnO$_2$-based sensors to several gases. Gas concentration: H$_2$ 8000 ppm; CH$_4$ 5000 ppm; C$_3$H$_8$ 2000 ppm; CO 200 ppm. Metal loading: 0.51 wt% for each. (From Yamazoe, N. et al., *Sensors Actuators, 4*, 283, 1983. With permission.)

REFERENCES

1. Taguchi, N., Japanese patent S47-38840.
2. Shaver, P. J., Activated tungsten gas detectors, *Appl. Phys. Lett.*, 11, 255, 1967.
3. Kul'kora, N. V. et al., The exchange of oxygen isotopes between carbon monoxide and carbon dioxide on an iron catalyst, *Dokl. Akad. Nauk* SSSR, 90, 1067, 1953.
4. Ihokura, K., Tin oxide gas sensors for deoxydizing gas, *New Mater. New Proc. Elecrochem. Technol.*, 1, 43, 1981.
5. Bond, G. C. et al., Oxidation of carbon monoxide catalyzed by palladium on tin IV oxide; an example of spillover, Proc. 6th Int. Congr. Catalysis, 1976, in Pub. 1977, The Chemical Society, London, 1976, 356.
6. Yamazoe, N., New approaches for improving gas sensors, Proc. 3rd Int. Meet. Chem. Sensors, Cleveland, 1990, 3.
7. Matsushima, S. et al., TEM observation of the dispersion state of Pd on SnO$_2$, *Chem. Lett.*, C.S.J. 1989, 1651, 1989.
8. Yamazoe, N. et al., Effects of additives on semiconductor gas sensors, *Sensors Actuators,* 4, 283, 1983.
9. Ohno, N., Japanese patent, S48-90793.
10. Taguchi, N., Japanese patent, S50-30480.
11. Maekawa, T. et al., H$_2$S detection using a CuO - SnO$_2$ sensor, Dig. 11th Chem. Sensor Symp., *Chem. Sensor,* 6, Supp. B21, 1990.
12. Matsushima, S. et al., Role of additives on alcohol sensing by semiconductor gas sensors, Chem. Lett., 845, 1989.
13. Morrison, S. R., Semiconducting oxide chemical sensors, *IEEE Circuits Syst.,* 7, 32, 1991.

CHAPTER FOUR

Selectivity

4.1 THE PROBLEM OF SELECTIVE BEHAVIOR

The most desirable sensor would be one which is sensitive to only a single gas and is not affected by others at all. However, because the sensing mechanism of stannic oxide is based on the catalytic oxidation of reducing gases on its surface, it is extremely difficult to confer a major degree of selectivity to such a system, for any reducing gas in the atmosphere will inevitably be detected. However, it is possible to enhance the sensitivities to certain groups of gases compared with those to other groups and to do so sufficiently for practical application.

A number of manufacturing and operating parameters has already been seen to affect the relative sensitivity to various gases. However, in this context, changes in crystallite size or the amount of binder have also been seen to affect other characteristics, such as long-term stability, so that modifications to these (already optimized) parameters are rarely allowable. Conversely, the proper choice of additives (such as palladium) and the quantities incorporated, and also the sensor operating temperature, can modify the relative sensitivities considerably without adversely affecting other parameters, including the long-term reliability.

In this chapter, the realization of various degrees of selectivity by operating temperature selection, and the appropriate choice of type and amount of additives, will be considered.

4.1.1 The Control of Relative Sensitivity Via Sensor Operating Temperature

Figure 4.1 gives the temperature characteristics of a sensor containing 0.2% of palladium to 1000 ppm of various gases in air and to clean air itself. Here, it will be seen that at 600°C, the sensor resistance is lowest for hydrogen, rising in the order ethanol, isobutane, methane, and carbon monoxide. That is, the gas sensitivity is greatest for hydrogen and falls in the order given above.

However, at 400°C, the greatest gas sensitivity is to ethanol, followed by hydrogen, carbon monoxide, isobutane, and finally methane, in descending order.

For these two temperatures, gas concentration characteristics have been plotted in Figures. 4.2 and 4.3, where the datum resistance R_o has been taken as R_a (that is, the reciprocal of the gas sensitivity has been plotted). These

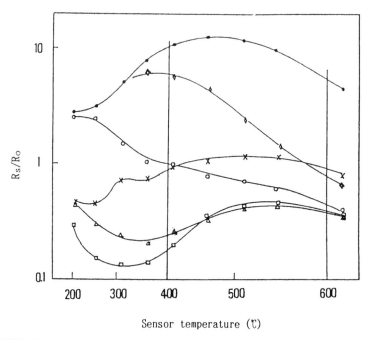

Sensor temperature (℃)

FIGURE 4.1   Sensor temperature characteristics for a sensor with 0.2% added palladium. R_S: sensor resistance for each gas at various sensor temperatures; R_o: 0.48 kΩ (R_S for 1000 ppm isobutane at sensor temperature 450°C); gas concentration: 1000 ppm. (●): Air; (○): isobutane; (△): hydrogen; (□): ethanol; (×): carbon monoxide; (◇): methane.

graphs clearly show that gas sensitivity is a function of operating temperature, as is the response speed, which was presented in the same format in Figure 2.7.

4.1.2   The Control of Relative Sensitivity Using Additives

Another set of temperature characteristics has been plotted in Figure 4.4 for a sensor with 2.0% of added palladium, and again in 1000 ppm of various gases. At 400°C this sensor is seen to be most sensitive to hydrogen, followed by isobutane, ethanol, methane, and carbon monoxide in that order. The gas concentration plot for a 400°C working temperature is shown in Figure 4.5, and this is seen to be very different from the plot of Figure 4.3, so demonstrating that the relative gas sensitivity can, in principle, be modified by controlling the amounts of additive incorporated.

Both platinum and gold may be added using the same techniques as described for palladium in Section 3.1.1, and the sensor temperature characteristics of Figures. 4.6 and 4.7 show the effects of including 2.0% of each, respectively.

For a sensor with 2.0% added palladium and operated at 400°C, Figure 4.4 showed that the sensitivity to carbon monoxide was poorer than those to all the other gases in the test group. However, for similar percentages of added platinum and gold, respectively, sensors operating again at 400°C display much improved CO-sensitivities, as shown by Figures. 4.6 and 4.7. In fact, for sensors with added

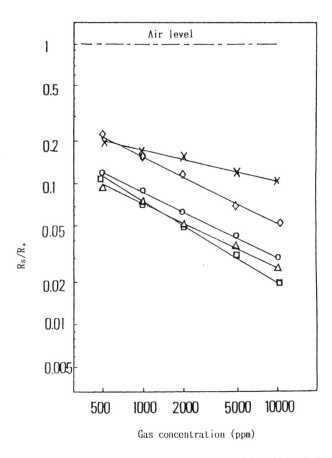

FIGURE 4.2 Gas concentration characteristics for a sensor with 0.2% added palladium. Sensor temperature: 600°C; R_S: sensor resistance for each gas at various concentrations; R_a: sensor resistance in air. (○): Isobutane; (△): hydrogen; (□): ethanol; (×): carbon monoxide; (◇): methane.

gold (Figure 4.7), the CO-sensitivity is greater than that to either methane or isobutane at 400°C. This demonstrates the possibility of sensitivity modification by the use of specific additives. However, attention must always be paid to concomitant side effects, such as those relevant to response speed and initial stabilization time, as demonstrated by Figures 3.7 and 3.8.

It may, therefore, be validly concluded that by the appropriate choice of sensor operating temperature, and the type and amount of additive, considerable practical improvements may be made with respect to the relative sensitivities of stannic oxide-based gas sensors.

4.2 SELECTIVITY TO CARBON MONOXIDE

The need for carbon monoxide selectivity is exemplified by areas where piped gas is used in commercial cooking operations. If town gas (consisting

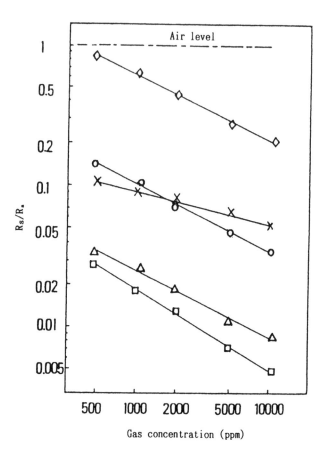

Gas concentration (ppm)

FIGURE 4.3   Gas concentration characteristics for a sensor with 0.2% added palladium. Sensor temperature: 400°C; R_S: sensor resistance for each gas at various concentrations; R_a: sensor resistance in air. (○): Isobutane; (△): hydrogen; (□): ethanol; (×): carbon monoxide; (◇): methane.

largely of carbon monoxide and hydrogen) is used, there can obviously be a build-up of these gases; if natural gas (methane) is used, incomplete combustion can also lead to the presence of carbon monoxide and hydrogen. Having regard to the toxicity of carbon monoxide, one specification called for its detection at the 200 ppm level in the presence of 1000 ppm of both ethanol vapor and hydrogen. (The required lack of sensitivity to ethanol recognizes the possible presence of alcohol vapor in some recipes.)

A combination of additive techniques and temperature control resulted in a sensor (now the Figaro TGS 203) suitable for this purpose. This sensor contains 0.5% palladium and 0.5% platinum, and its working temperature is 80°C. The relevant temperature characteristics, Figure 4.8, clearly indicate that its sensitivity to carbon monoxide is indeed greater than those to either hydrogen or ethanol vapor at 80°C.

A gas concentration characteristic for such a sensor working at 80°C is given in Figure 4.9, and here it is seen that the concentration of hydrogen

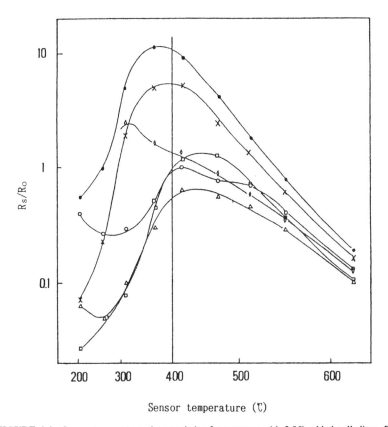

Sensor temperature (°C)

FIGURE 4.4 Sensor temperature characteristics for a sensor with 2.0% added palladium. R_S: sensor resistance for each gas at various sensor temperatures; R_o: 18.60 kΩ (R_S for 1000 ppm isobutane at sensor temperature 450°C); gas concentration: 1000 ppm. (●): Air; (○): isobutane; (△): hydrogen; (□): ethanol; (×): carbon monoxide; (◇): methane.

reached 2300 ppm, or that of ethanol vapor reached 3000 ppm, before the sensor resistance equalled that at 200 ppm of carbon monoxide (which was a very low 80.6 Ω for the sensor sample involved). Thus, an appropriate alarm circuit would operate at 200 ppm of carbon monoxide, but would remain quiescent for the other gases until the much higher levels quoted were reached.

Unfortunately, at the very low working temperature of 80°C, the stannic oxide sensor is unstable, and its resistance rises gradually with time, as shown by the dotted line in Figure 4.10. This is considered to result from the facts that both atmospheric water vapor and organic components with high boiling points adsorb easily, but desorb very little around 80°C, so that the resultant resistance changes are cumulative and so adversely affect the long-term stability of the sensor.

To obviate these effects, it is necessary to heat clean the surface periodically, and in the commercial TGS 203, this is done automatically by cyclically heating the sensor to 260°C for 1 minute, then stabilizing it to 80°C over 1.5 minutes, and finally measuring its resistance quickly, after which the

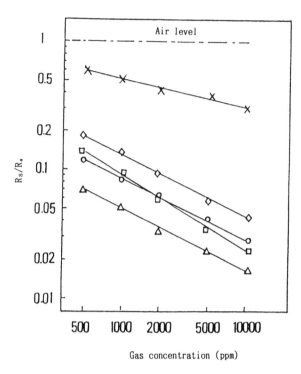

Gas concentration (ppm)

FIGURE 4.5   Gas concentration characteristics for a sensor with 2.0% added palladium. Sensor temperature: 400°C; R_S: sensor resistance for each gas at various concentrations; R_a: sensor resistance in air. (O): Isobutane; (Δ): hydrogen; (□): ethanol; (×): carbon monoxide; (◇): methane.

cycle repeats. When this mode of operation is employed, good long-term performance is achieved, as is shown by the solid line in Figure 4.10.

4.3   SELECTIVITY TO AMMONIA

Sensors selective to ammonia have numerous applications, one of the most important being monitoring of the working environment in refrigerating plants for protection of the operatives. The TLV (tolerance limit value for a working shift of 8 hours per day) for ammonia is currently 25 to 50 ppm in most countries, which means that the relevant sensor must be sufficiently sensitive and selective to cover such low concentrations.

Empirical work involving various additives and working temperatures has shown that the optimum conditions for ammonia selectivity are 1% of palladium and 300°C. This latter value is confirmed by the sensor temperature characteristics of Figure 4.11.

Figure 4.12 is a gas concentration characteristic diagram which compares the response of sensors with similar palladium concentrations (1%) to both ammonia and methane at working temperatures of 300 and 450°C, respectively. It clearly shows that the relative sensitivity for the ammonia sensor at its

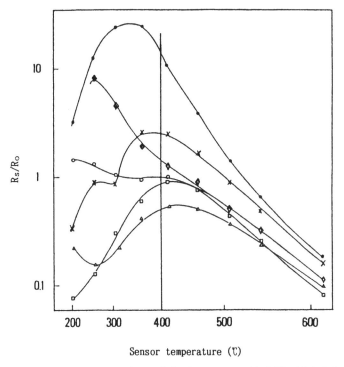

Sensor temperature (℃)

FIGURE 4.6 Sensor temperature characteristics for a sensor with 2.0% added platinum. R_S: sensor resistance for each gas at various sensor temperatures; R_o: 26.76 kΩ (R_S for 1000 ppm isobutane at sensor temperature 450°C); gas concentration: 1000 ppm. (●): Air; (○): isobutane; (△): hydrogen; (□): ethanol; (×): carbon monoxide; (◇): methane.

optimum temperature of 300°C is more than adequate for practical applications which necessitate distinguishing between these two gases.

4.4 SELECTIVITY TO ETHANOL VAPOR

There are also numerous applications of a sensor which is selective to ethanol, ranging from process control of alcohol fermentation to the measurement of expired alcohol in the breath. This is a particularly easy requirement for stannic oxide sensors because without additives of any sort, they respond preferentially to ethanol (and hydrogen) as compared with their rather low general sensitivity. This is clearly seen in Figure 3.1, where particularly good ethanol sensitivity is observed at a working temperature around 200°C. Figure 4.13 gives the relevant gas concentration characteristics and again illustrates the marked sensitivity to ethanol of a sensor without additives.

4.5 RECENT WORK ON SELECTIVITY

Many workers have vigorously pursued research into methods of realizing selective solid-state gas sensors, concentrating on four factors, three of which

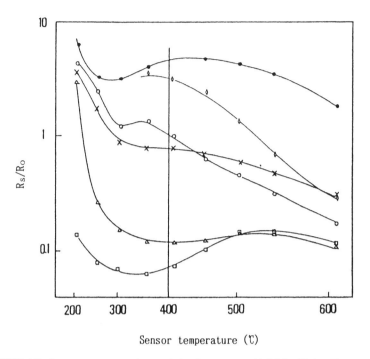

Sensor temperature (℃)

FIGURE 4.7 Sensor temperature characteristics for a sensor with 2.0% added gold. R_S: sensor resistance for each gas at various sensor temperatures; R_o: 0.83 kΩ (R_S for 1000 ppm isobutane at sensor temperature 450°C); gas concentration: 1000 ppm. (●): Air; (○): isobutane; (△): hydrogen; (□): ethanol; (×): carbon monoxide; (◇): methane.

have already been described in the present chapter: to wit, type and amount of additive and working temperature. The fourth may be described as the "form factor", or geometry of the sensor, but this has not yet been considered herein. So far, a rational methodology for the "tailoring" of selectivity has not emerged, but Table 4.1 lists some papers using various approaches reported from 1983 to 1992.

As mentioned earlier, Yamazoe et al. have proposed that in searching for appropriate additives, not only should their catalytic activity be taken into account, but so should their receptor function for gas detection. This group has published extensively on selectivity to both alcohol vapor and hydrogen sulfide, and their reports on the latter will be summarized in the following section.

4.5.1 Selectivity to Hydrogen Sulfide

The addition of various metal ions to stannic oxide sensor material has been investigated by Maekawa and Yamazoe et al.[1] in terms of hydrogen sulfide selectivity, and their conclusion is that the sensitivity increases with decreasing electronegativity (that is, increasing basicity) of the additive. However, a sensor with added (bivalent) copper ions showed a peculiarly high sensitivity to hydrogen sulfide, as is seen in Figure 4.14.

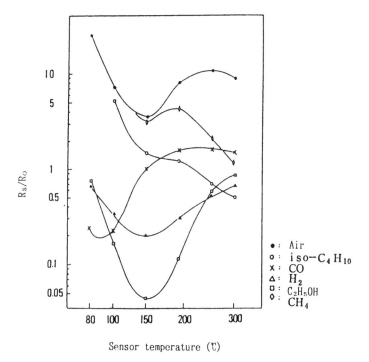

Sensor temperature (℃)

FIGURE 4.8 Sensor temperature characteristics of the stannic oxide sensor for carbon monoxide. Additives: palladium 0.5%, platinum. 0.5%; R_s: sensor resistance for each gas at various sensor temperatures; R_o: sensor resistance for 500 ppm carbon monoxide at sensor temperature 150°C); gas concentration: 500 ppm.

The relevant sensor temperature characteristic is shown in Figure 4.15, from which it was concluded that the optimum response to hydrogen sulfide occurred near 200°C for 3 wt% to 5 wt% of added copper oxide (which is the form in which the copper ions exist on the stannic oxide surface).

Table 4.2 shows the remarkable sensitivity of the sensor with added copper to hydrogen sulfide as compared with several other gases,[1] and the mechanism underlying this effect may be as follows.

Knowing that the copper oxide particles existing on the sensor surface are p-type semiconductors, they must form pn junctions with the stannic oxide surface. Therefore, the electron density in the surface layers of the stannic oxide must be low, which accounts for the high sensor resistance in air.

In the presence of hydrogen sulfide, the copper oxide converts to the sulfide at 200°C or less (according to Reaction 4.1 below), the electron density increases, and the resistance of the sensor falls. Recovery in clean air conforms to Reaction 4.2 when the sensor returns to its original high resistance state.

$$CuO + H_2S \rightarrow CuS + H_2O \qquad (4.1)$$

$$CuS + {}^3/_2 O_2 \rightarrow CuO + SO_2 \qquad (4.2)$$

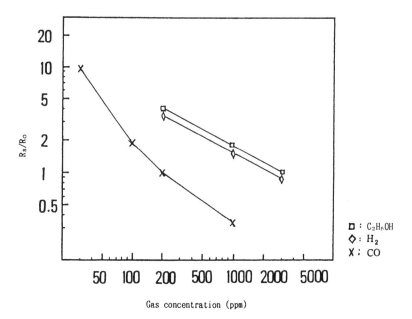

FIGURE 4.9   Gas concentration characteristics of the stannic oxide sensor for carbon monoxide. Additives: palladium 0.5%, platinum 0.5%; sensor temperature: 80°C when detecting carbon monoxide; R_S: sensor resistance for each gas at various concentrations; R_o: 80.6 kΩ (R_S for 200 ppm carbon monoxide).

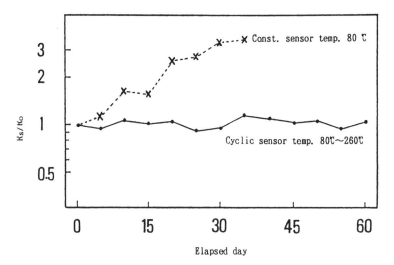

FIGURE 4.10   Comparison of a carbon monoxide sensor operated continuously at 80°C (dotted line) and with 260°C cyclical heat cleaning. R_S: sensor resistance for 200 ppm carbon monoxide at each measurement; R_o: the first measured value of R_S; cycle of sensor temperature: 80°C for 90 sec; 260°C for 60 sec; measurement done at the end of 80°C period.

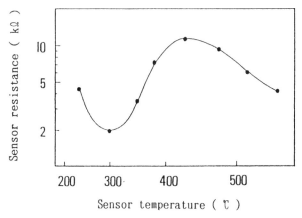

FIGURE 4.11 Sensor temperature characteristic of a stannic oxide sensor with 1.0% added palladium in 1000 ppm ammonia gas.

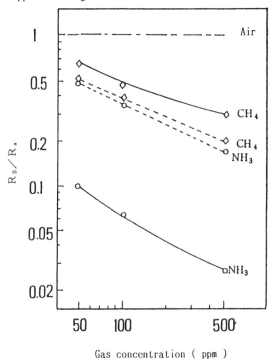

FIGURE 4.12 Gas concentration characteristics of an ammonia-selective sensor compared with those of a general purpose sensor. (——): The ammonia-selective sensor (sensor temperature: 300°C, added palladium: 1.0%); (------): the sensor for general combustible gases (sensor temperature: 450°C, added palladium: 1.0%). R_s: sensor resistance in gas; R_a: sensor resistance in air.

The sulfuration reaction represented by Equation (4.1) plays the dominant role in the gas-sensing mechanism described above because the reaction is unique to hydrogen sulfide, resulting in the copper oxide functioning as a hydrogen sulfide receptor, and so conferring selectivity to this gas upon the sensor.

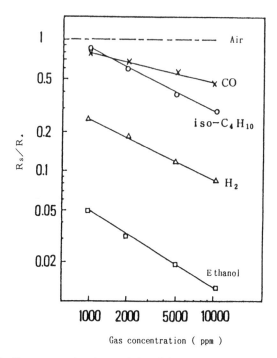

FIGURE 4.13   Gas concentration characteristics of the ethanol-selective sensor. Added palladium: none; sensor temperature: 200°C.

TABLE 4.1   **Short List of Research Papers on Selectivity in Stannic Oxide Gas Sensors, 1983/92**

1. Wada, K., Effect of additives on the gas sensing characteristics of semiconducting metal oxide gas sensors. 1 – platinum-stannic oxide & palladium-stannic oxide, Sasabo Kogyo Koto Senmon Gakko enkyu Hokoku, 20, 37, 1983.
2. Wang, C. *et al.,* Research & applications of gas sensing semiconductors, *Anal. Chem. Symp. Ser.,* 17, (Chem. Sens), 7, 1983.
3. Lalauze, R. *et al.,* A new approach to the selective detection of gas by a tin dioxide solid-state sensor, *Sensors Actuators* 5, 55, 1984.
4. Toshiba Corp., Gas Sensor, Japanese Patent 60007353, 1985.
5. Coles, G.S.V. *et al.,* Fabrication and preliminary tests on tin (IV) oxide-based gas sensors, *Sensors Actuators* 7, 89, 1985.
6. Fukui, K., A high-selectivity hydrogen sensor using a semiconductor with a hot-wire heater and its application, *Keiso,* 28, 56, 1985.
7. Jeon, B.S. *et al.,* Preparation of stannic oxide semiconducting gas sensor by wet process, *Yoop Hakhoechi,* 23(3), 53, 1986.
8. Wada, K. *et al.,* Detection of carbon monoxide using platinum tin oxide, Sasebo Kogyo Koto Senmon Gakko Kenkyu Hokoku, 23, 59, 1986.
9. Fleischmann, D. *et al.,* Manufacture of semiconductor gas sensor, German Patent DE 3529820, 1986.
10. Matsuo, K. *et al.,* Gas sensing properties of tin oxide sensors doped with noble metals, *Nagasaki Daigaku Kogakubu Kenkyu Hokoku,* 17, 67, 1987.
11. Morrison, R.S., Selectivity in semiconductor gas sensors, *Sensors Actuators,* 12, 425, 1987.

TABLE 4.1 (continued)   Short List of Research Papers on Selectivity in Stannic Oxide Gas Sensors, 1983/92

12. Shen, H. *et al.*, Sensing characteristics of cerium dioxide-doped tin dioxide carbon monoxide gas sensor, *Yingyong Kexue Xuebao*, 6, 175, 1988.
13. Dramlic, D.M., Determination of dominant reactions and the degree of doping in commercial tin dioxide-type gas sensors, *Naucno Tekh. Pregl.*, 38, 33, 1988.
14. Sears, W.M. *et al.*, Algorithms to improve the selectivity of thermally-cycled tin dioxide gas sensors, *Sensors Actuators*, 19, 333, 1989.
15. Demarne, V. *et al.*, Integrated semiconductor gas sensor evaluation with an automatic test system, *Sensors Actuators* B1, 87, 1990.
16. Weimar, U. *et al.*, Pattern recognition methods for gas mixture analysis: application to sensor arrays based upon SnO$_2$, *Sensors Actuators* B1, 93, 1990.
17. Sberveglieri, G. *et al.*, Response to nitric oxide of thick and thin SnO$_2$ films containing trivalent additives, *Sensors Actuators*, B1, 79, 1990.
18. Coles, G.S.V. *et al.*, Selectivity studies on tin oxide-based semiconductor gas sensors, *Sensors Actuators*, B3, 7, 1991.
19. Mattila, A.H. *et al.*, Sensitivity and selectivity of doped SnO$_2$ thick film sensors to H$_2$S in the constant and pulsed temperature modes, *Sensors Actuators*, B6, 248, 1992.
20. Hiranaka, Y., Abe T., and Murata, H., Gas-dependent response in the temperature transient of SnO$_2$ gas sensors, *Sensors Actuators*, B9, 177, 1992.

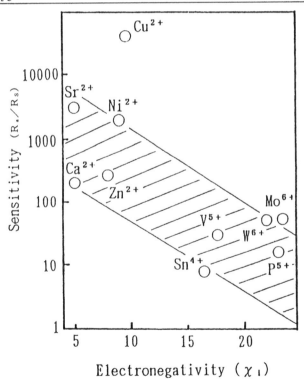

FIGURE 4.14   Sensitivity of 5 wt% metal-oxide-added SnO$_2$ sensors to 50 ppm H$_2$S at 200°C as a function of the electronegativities of added metal cations. (From Maekawa, T., Yamazoe, N. et al., Dig. 11th Chem. Sensor Symp., *Chem. Sensor*, 6, 21 (Suppl. B), 1990. With permission.)

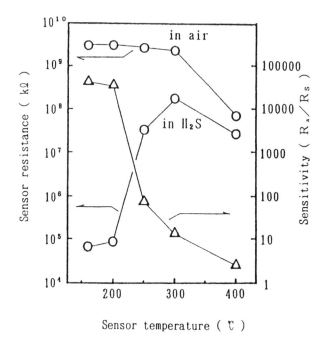

Sensor temperature (°C)

FIGURE 4.15 Sensor temperature characteristics and sensitivity to hydrogen sulfide of a stannic oxide gas sensor with 5 wt% of copper oxide. (From Maekawa, T., Yamazoe, N. et al., Dig. 11th Chem. Sensor Symp., *Chem. Sensor.*, 6, 21 (Suppl. B), 1990. With permission.)

TABLE 4.2   Comparison of Sensitivity of CuO (5 wt%)-SnO$_2$ Sensor to Various Gases

Gas	H$_2$S	CO	iso-C$_4$H$_{10}$	C$_2$H$_5$OH	H$_2$
Concentration (ppm)	50	1000	1000	1000	800
Sensitivity (R$_a$/R$_s$)	35,000	1.3	1.2	1.9	1.0

From Maekawa, T., Yamazoe, N. et al., Dig. 11th Chem. Sensor Symp., *Chem. Sensor.*, 6, (Suppl. B), 21, 1990.

4.5.2 Further Effects of Metal Oxide Additives

Other effects of metal oxide additives have been investigated by Matsushima and Yamazoe et al.[2] including those relevant to ethanol. As shown in Figure 4.16, they found that, in general, the addition of basic oxides enhanced ethanol sensitivity, whereas the addition of acid oxides degraded it. From this work, they concluded that ethanol sensitivity may be controlled by adjustment of the acidity or basicity of the sensor surface.

When ethanol is oxidized, either acetaldehyde or ethylene is formed as an intermediate step, and it has been reported that it is the former which takes place on the surface of a catalyst and the latter on an acidic surface. The observed high sensitivity of the stannic oxide sensor to acetaldehyde implies that it is this intermediary which is active during the oxidation of ethanol and which accounts

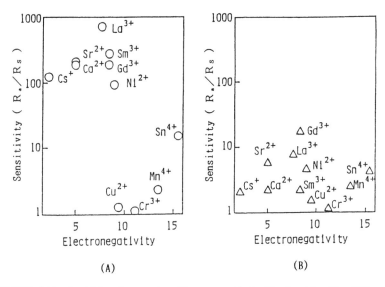

FIGURE 4.16 Sensitivity of a stannic oxide sensor with 5 wt% of metal oxide additives of various electronegativities. (A) In 1000 ppm of ethanol; (B) in 1000 ppm of isobutane; working temperature: 300°C. (From Matsushima, M., Yamazoe, N. et al., *Chem. Lett.*, 1989, 845, 1989. With permission.)

for the improvement in ethanol sensitivity accruing from the addition of basic oxides.

Of these oxides, it is that of lanthanum, La_2O_3, which gives rise to the greatest selectivity to ethanol and does so at 300°C, as shown in Figure 4.17.

It should be noted from Figure 4.16(B) that the addition of metal oxides appears to make little difference to the isobutane sensitivity, which illustrates a clear case of selectivity enhancement between these two gases.

4.6 MAJOR ADDITIVE TECHNIQUES

In some cases, the amount of additive utilized is so large that the mix might be considered to represent a new form of active material entirely. For example, Coles et al.[3] have reported that stannic oxide mixed with 36 wt% of aluminum silicate and also 1.5 wt% of palladium chloride ($PdCl_2$) results in a sensor which is selective to methane in the presence of carbon monoxide. An aqueous paste of this material is applied to an alumina substrate, dried at room temperature, then sintered at 1000°C for some 2 hours. (It should be noted that the electrodes and heater, which are deposited on opposite sides of the 0.5 mm thick substrate, must be of platinum rather than gold to avoid metal migration at such a high sintering temperature.)

The converse — selectivity to carbon monoxide in the pressure of methane — was achieved by mixing 15 wt% of bismuth oxide (Bi_2O_3) to the stannic oxide and sintering only at 800°C. Figure 4.18 is a normalized resistance plot for this latter sensor showing its selectivity to carbon monoxide and hydrogen

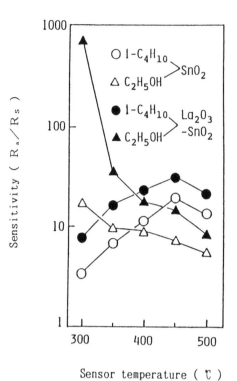

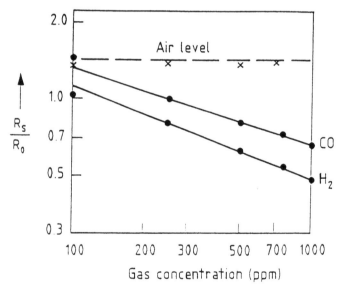

FIGURE 4.17 Temperature dependences of sensitivity of SnO_2 sensors with and without La_2O_3 to 1000 ppms of ethanol and isobutane. (From Matsushima, S., Yamazoe, N. et al., *Chem. Lett.,* 1989, 845, 1989. With permission.)

FIGURE 4.18 Resistance ratio characteristics for a SnO_2 - Bi_2O_3 sensor showing selectivity to carbon monoxide and hydrogen compared with methane. R_o = 37.3 mΩ at 250 ppm CO; x: response to CH_4.

compared with methane. It was noted that for this form of sensor, several methods of stannic oxide powder preparation gave similar results, and this was thought to be due to the fact that the particle sizes were similar (3.8 to 5.4 μm) in all cases, irrespective of wide variations in specific surface area.

The further enhancement of selectivity to hydrogen for mixed oxide sensors has also been reported,[4] but, unfortunately, sensors like these do have the disadvantage of exhibiting very high resistances in air — up to 50 mΩ in fact. However, Coles et al.[4] have shown that these resistances may be reduced to more easily measurable levels by adding about 2 wt% of antimony oxide (Sb_2O_3) .

The effects of antimony additives has been investigated comprehensively,[5-7] and the general *modus operandi* appears to be modification of the stannic oxide work function via valency control.

However, the operating mechanisms of the bismuth oxide is not understood, but must be quite different from that of the alumina described in Chapter 1, which depended on purely physical phenomena to enhance both the strength and the sensitivity of the ceramic.[8]

REFERENCES

1. Maekawa, T., Yamazoe, N., et al., H_2S detection using CuO-SnO_2 sensor, Dig. 11th Chem. Sensor Symp., *Chem. Sensor* (E.S.J.), 6, (Supp. B) 21, 1990.
2. Matsushima, S., Yamazoe, N., et al., Rôle of additives on alcohol sensing by semiconductor gas sensor, *Chem. Lett.,* (C.S.J.)1989, 845, 1989.
3. Coles, G.S.V., Gallagher, K.J., and Watson, J., Fabrication and preliminary tests on tin (IV) oxide-based gas sensors, *Sensors Actuators,* 7, 89, 1985.
4. Coles, G.S.V., Williams, G., and Smith, B., Selectivity studies on tin oxide-based gas sensors, *Sensors Actuators,* B3, 7, 1991.
5. Ohno, N., Japanese Patent S48-90793.
6. Uematsu, K., Mizutani, N., and Kato, M., Electrical properties of high purity tin oxide doped with antimony, *J. Mater. Sci.,* 22, 915, 1987.
7. Yannopoulos, L.N., Antimony-doped stannic oxide-based thick film gas sensors, *Sensors Actuators,* 12, 77, 1897.
8. Taguchi, N., Japanese Patent 550-30480.

CHAPTER 5

The Stannic Oxide Gas Sensor as a Combustion Monitor

5.1 METHODS OF MONITORING COMBUSTION PRODUCTS

The monitoring of combustion products is of continuing technological importance in terms of combustion efficiency, safety, and pollution control. While maximal thermal efficiency should be maintained on grounds of fuel economy, safety factors must always predominate, and this is well-illustrated in the cases of both automobile engines and certain forms of domestic heater, for example, both of which can produce copious quantities of highly toxic carbon monoxide. (At the time of writing, the tolerance limit value or TLV level for carbon monoxide is 35 ppm in the U.S. and 50 ppm in Japan and Europe.) Furthermore, the internal combustion engine is recognized as a major source of pollution, some of which results from incomplete combustion.

The most common form of solid-state combustion sensor is actually the zirconia oxygen sensor,[1] including the more recent oxygen-pumping sensor,[2] but both titania and cobalt oxide have also been evaluated for this purpose.[3,4] Unfortunately, the zirconia sensor remains a somewhat expensive device, while the titania sensor has some long-term stability problems, and the cobalt oxide sensor produces too small a signal for general usage.

The use of an oxygen sensor depends upon the fact that, given an oversupply of air, some free oxygen will be present in the exhaust gas mix (lean burning) while if insufficient air for complete combustion is present, there will be unburnt fuel in the exhaust gas mix (rich burning). For lean burning, where some oxygen is present in the exhaust stream, a stannic oxide sensor should present a high resistance for reasons previously explained in detail. For incomplete combustion, however, the resistance should fall as both surface and lattice oxygen is desorbed by the reducing power of combustible gases in the exhaust stream. Hence, the stannic oxide sensor should be capable of determining which side of the ideal or stoichiometric point combustion is occurring.

The development of such a device is described below and has led to commercialization in the form of the Figaro CMS-302 combustion sensor.[5,6]

5.2   FABRICATION OF THE COMBUSTION SENSOR

For installation in exhaust gases, or indeed in or near a flame, a combustion sensor will always be required to operate in a hot environment, often around 900°C, but which may easily exceed 1000°C. Hence, the entire structure must be designed to withstand such temperatures, but by the same token, a heater filament will obviously not be required. This is, in one sense, advantageous, but the drawback is that the sensor will be required to work within a very wide range of operating temperatures.

It has been found that an appropriate form of stannic oxide for use under these conditions may be obtained from dried tin hydroxide by calcination at 500°C for 4 hours. The resultant material is then ground and mixed with α-alumina of average grain size 28 μm in a 2:1 weight ratio. This mixture is then used without additives.

Figure 5.1 shows the construction of the sensor. Platinum electrodes are threaded through small-diameter holes in the alumina substrate, and the active material is pressed into the cavity shown. The whole structure is then inserted into an electric oven and sintered at 1400°C for 5 hours (which implies that it will easily withstand any normally encountered high-temperature environment). After sintering, lead wires are connected to the platinum electrodes to complete the assembly.

It should be noted here that although such sensors have been evaluated for use in automobile exhausts,[7] they will not respond rapidly until heated to several hundred degrees — that is, they are not useful immediately after starting an engine. However, some sensors include heaters to solve this problem, an example being the titania sensor shown in Figure 5.2.[8]

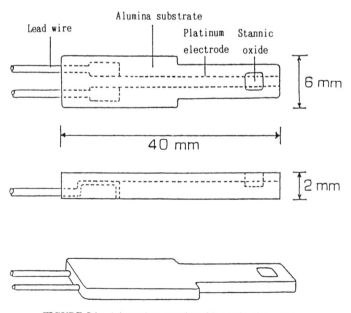

FIGURE 5.1   A heaterless stannic oxide combustion sensor.

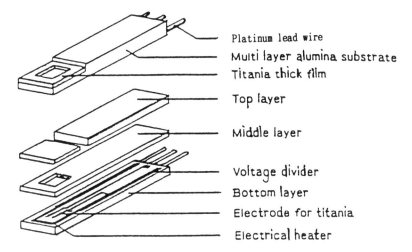

Platinum lead wire

Multi layer alumina substrate

Titania thick film

Top layer

Middle layer

Voltage divider

Bottom layer

Electrode for titania

Electrical heater

FIGURE 5.2 A titania combustion sensor with heater. (Reproduced with permission from Takami, A., et al., SAE Tech. Paper Series 870290. ©1987 Soc. Autom. Eng. Inc.)

5.2.1 A Combustion Sensor Evaluation Method

Figure 5.3 shows the layout of apparatus suitable for producing exhaust gases under controlled conditions. Here, flows of air and fuel gas (methane in this case) are defined using flow control valves, and both are mixed in a combustion furnace containing an oxidation catalyst consisting of alumina-supported palladium. Note that these parts of the system are duplicated, so that two different gas mixtures can be produced concurrently, leading to two different exhaust streams which can each be controlled to represent rich through the stoichiometric

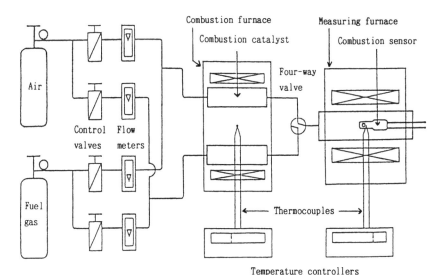

FIGURE 5.3 Measuring system for combustion sensor characterization.

point to lean combustion. However, both streams are at the same temperature, having been generated in the same furnace.

A four-way valve directs the exhaust gases into a measuring furnace so that the response speed of a sensor contained therein can be measured by switching quickly from one stream to the other. The sensor temperature is monitored by a thermocouple, as shown.

The excess air ratio λ of the fuel-air mix is defined so that for a stoichiometric induction mix, $\lambda = 1$, whereas for lean mixtures $\lambda > 1$ and for rich mixtures $\lambda < 1$. This ratio therefore provides a convenient base line for the characterization of combustion sensors as detailed below. Here, the measuring circuit used was that of a series load, as shown in Figure 1.36, and the voltage applied across the series sensor-load combination was 1 V as further depicted in Figures 5.9 and 5.10.

5.2.2 Measured Sensor Characteristics

Figure 5.4 gives a plot of sensor resistance vs. the excess air parameter λ for a sensor working at three different temperatures, where the word "temperature"

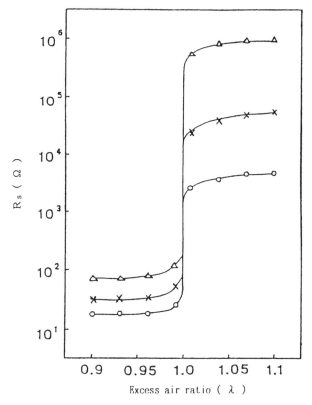

FIGURE 5.4 The relationship between sensor resistance and excess air ratio for a stannic oxide combustion sensor. Fuel gas: methane; sensor and exhaust gas temperature: (—○—) 800°C, (—×—) 600°C, (—△—) 400°C.

is taken to mean the exhaust gas temperature (which is also the temperature at which the sensor operates) for the purposes of this chapter. The relevant fuel gas was methane, and an artificial air mix of nitrogen and oxygen was also employed. This system was also used for gathering the further data presented in this chapter.

The most prominent feature of the graph is the sudden change in sensor resistance at $\lambda = 1$, that is, where excess fuel in the induction mix changes to excess air. Also it will be noted that this large resistance change occurs at $\lambda = 1$ irrespective of temperature within the ranges chosen, which are 400, 600, and 800°C.

From an applications viewpoint, the conclusion is that this form of sensor may be utilized to detect any departure from a stoichiometric induction mix and so may be used for control purposes where this mix is regarded as optimum. Such control must therefore be of a discrete nature because of the (intentional) steepness of the slope of the sensor characteristic at the stoichiometric point. This occurs because the sensor responds largely to the rapid change in equilibrium adsorbed oxygen density at $\lambda = 1$ (its surface being highly catalytically active) as opposed to the much more gradual change in ambient oxygen partial pressure. However, it has been suggested that for internal combustion engine exhaust applications, this sudden change in sensor resistance at the stoichiometric point[7] should be avoided in favor of a more gradual response. This would make continuous control potentially feasible, which is desirable because lean mixture settings are preferred in many cases for reasons of fuel economy and pollution reduction.

5.2.3 Temperature Dependency

Away from stoichiometry, such as at $\lambda = 0.9$ or $\lambda = 1.1$, Figure 5.4 shows that the sensor is very temperature dependent. Knowing that exhaust gas temperatures vary widely depending upon the conditions of combustion (and loading in the case of engines), it is clear that the temperature-dependent portions of the characteristic must be taken into account in practical applications. From this figure, it is obvious that the resistance change over the $\lambda = 1$ point is least from that part of the characteristic at $\lambda < 1$ and 400°C (at the left), to the $\lambda > 1$ and 800°C part (to the right). If this represents a resistance ratio of more than about 10:1 (which it does in this figure), then the sensor can be regarded as useful because appropriate electronics can easily recognize the transition over the stoichiometric point even taking into account random drift. Also, it is clear that the wider the difference between the two temperatures which result in this minimal acceptable resistance ratio, the greater the temperature span over which the sensor remains useful. This will certainly encompass ranges as wide as 400 to 900°C.

The temperature dependency may be replotted from Figure 5.4 at two values of λ (0.9 and 1.1) to give the graph of Figure 5.5, and from this the resistance ratio may be extracted for any pair of temperatures.

It can be concluded that the stannic oxide sensor could be competitive with zirconia or titania sensors, especially when it is recalled that the latter devices incorporate catalytic material (usually platinum in the case of zirconia) which is not only expensive, but which eventually deteriorates.

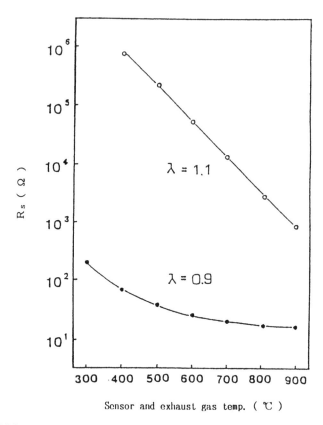

Sensor and exhaust gas temp. (℃)

FIGURE 5.5   Temperature dependency of a stannic oxide combustion sensor. Fuel gas: methane; (—○—): sensor resistance at excess air ratio 1.1; (—●—): sensor resistance at excess air ratio 0.9.

Figure 5.5 may also be replotted to give graphs of sensor conductivity vs. the reciprocal of the (absolute) exhaust temperature, as is shown in Figure 5.6. This presentation indicates that the slope of the sensor temperature dependency changes near 600°C for $\lambda = 1.1$, where enough oxygen remains in the exhaust gases for adsorption onto the stannic oxide surface to take place. This implies that two donors having different ionization energies must contribute to the semiconducting properties of the stannic oxide.

If the sensor is exposed only to an oxygen-nitrogen mix at high temperatures, characteristics such as those of Figure 5.7 result. For both of the plots shown, the law followed was as given in Section 2.3.1:

$$G_S \propto P_{O_2}^{-\beta} \tag{5.1}$$

Again, the value of β ought to be 0.5 if G_S were dependent only upon the dissociative adsorption of oxygen and 0.17 if it were defined only by lattice oxygen defects.[9] However, the values of β derived from Figure 5.7 are 0.40 at 400°C and 0.19 at 900°C.

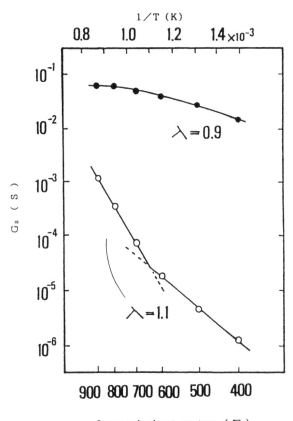

FIGURE 5.6 Conductivity vs. 1/T for a stannic oxide combustion sensor. Fuel gas: methane; (—○—): sensor conductivity at excess air ratio 1.1; (—●—): sensor conductivity at excess air ratio 0.9.

According to Yamazoe et al.,[10] oxygen which desorbs from a stannic oxide surface at 400 to 600°C is that adsorbed dissociatively in the form of O^- or O^{2-}, while the desorption of lattice oxygen occurs at 600°C or more. Hence, it can be assumed that the sensor conductivity for complete combustion in the $\lambda = 1.1$ region is defined mainly by the density of oxygen adsorbed dissociatively onto the stannic oxide surface at 600°C or less, as was the case for the sensor materials discussed earlier. At higher temperatures, the conductivity is thought to be defined largely by the bulk lattice oxygen defect density, which can equilibrate with the oxygen partial pressure at these high temperatures.

These concepts can be used to explain two phenomena illustrated in Figure 5.6. Firstly, the sensor conductance G_s at $\lambda = 0.9$ is much higher than at $\lambda = 1.1$; secondly, its dependence on temperature at $\lambda = 0.9$ is smaller than at $\lambda = 1.1$ and does not exhibit the clear transition from one slope to another seen in the latter case. These phenomena may be explained as follows.

Regarding the absolute value of G_s, when $\lambda = 0.9$ the oxygen partial pressure is very low — some 10^{-15} torr or less[10] — so that the oxygen lattice defect density becomes large and, hence, so does the conductivity. However,

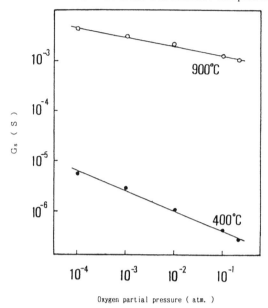

FIGURE 5.7   Sensor conductivity vs. oxygen partial pressure for a stannic oxide combustion sensor. Balance gas: nitrogen; (—○—): sensor conductivity at 900°C; (—●—): sensor conductivity at 400°C.

at the lower temperatures, the effect of the reducing gas component in the exhaust stream becomes more pronounced, that is, the difference between the values of G_S at $\lambda = 0.9$ and $\lambda = 1.1$ becomes larger. To explain this, consider the following experiment. If exhaust gas at $\lambda = 0.9$ is diluted with nitrogen, and the sensor conductance is plotted against the dilution ratio (defined as the gas to nitrogen ratio by volume), Figure 5.8 shows that at 900°C, the sensor conductance is high and almost constant. This implies that the very low oxygen partial pressure alone defines G_S. This is reasonable if it is recalled that at such a high temperature, the surface adsorbed oxygen density will be very low, so that the effect of the interaction between any reducing gas and this surface oxygen will be minimal. So, only the effect of the low partial-pressure atmospheric oxygen on the bulk conductivity will be seen. However, at 400°C, G_S rises with the dilution ratio, which implies that at comparatively low temperatures, the interaction between the surface adsorbed oxygen and the reducing gas component becomes significant. In short, oxygen partial pressure alone defines G_S at higher temperatures (such as 900°C), but at lower temperatures (such as 400°C) the reducing gas component also contributes.

Returning to Figure 5.6, if G_S were defined only by oxygen partial pressure, then its values at $\lambda = 0.9$ and 1.1 would differ by the same amount at 400°C as at 900°C, and the curves at the two values of λ would be essentially parallel. However, at low temperatures (such as 400°C) the (reducing) exhaust gas at $\lambda = 0.9$ removes the adsorbed oxygen from the sensor surface, and this effect is much greater than that due to oxygen partial pressure, for the lattice oxygen hardly desorbs at all at these temperatures, as has been pointed out.[10] This explains the smaller temperature dependence of G_S at $\lambda = 0.9$, for the density of surface oxygen decreases as the temperature rises, so the effect of its

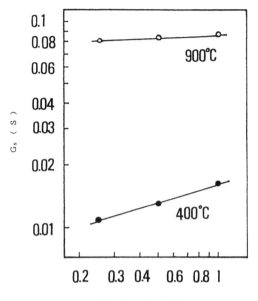

Dilution ratio of exhaust gas.

FIGURE 5.8 Sensor conductivity vs. exhaust gas concentration for a stannic oxide combustion sensor. Exhaust gas: from combustion of methane at excess air ratio 0.9; diluent: nitrogen; (—○—): sensor conductivity at 900°C; (—●—): sensor conductivity at 400°C.

consumption by the reducing gas component decreases until G_S is defined only by the partial pressure of oxygen in the exhaust gas. This process becomes complete by about 900°C in spite of the continued presence of reducing gas, the result being this smaller dependence of G_S on temperature at $\lambda = 0.9$ clearly seen in the upper plot of Figure 5.6.

The reason why the transition seen at $\lambda = 1.1$ in Figure 5.6 is not observed at $\lambda = 0.9$ may well be that the marked effect of the reducing gas present at $\lambda = 0.9$ conceals the change in the temperature coefficient of conductivity resulting from the underlying mechanisms described above.

5.2.4 Response Speed

When the four-way valve in the test system of Figure 5.3 is switched rapidly to inject exhaust gas streams of $\lambda = 1.1$ and $\lambda = 0.9$ alternately into the measuring furnace, the result is shown in Figure 5.9. Here, as an indication of practical utility, the voltage across the stannic oxide sensor load resistor has been measured and compared with the e.m.f. generated by a stabilized zirconia sensor. This comparison is therefore a good demonstration of the equal response speeds of the two forms of sensor in terms of practical operation.

5.2.5 Durability

To determine consistency of operation of the stannic oxide combustion sensor over a long period, the exhaust stream was switched from $\lambda = 1.1$ to $\lambda = 0.9$

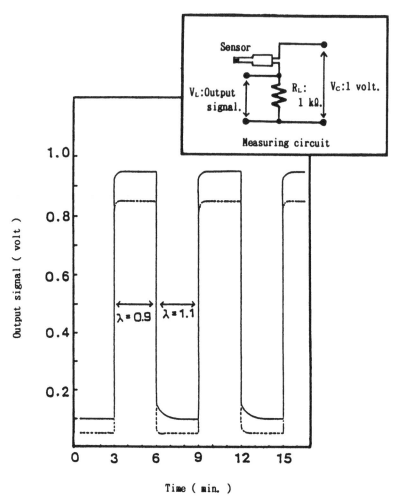

FIGURE 5.9   Response speeds of stannic oxide and zirconia sensors. Fuel gas: methane; sensor and exhaust gas temperature: 600°C; (———): V_L of stannic oxide sensor; (------): electromotive force of zirconia oxygen sensor.

every 3 minutes over some 3 months, which is about 22,000 times. Figure 5.10 shows a number of measurements taken during this process over the entire period, and it will be seen that there was no observable long-term deterioration.

A further test of over 12,000 hours of continuous use at 900°C and $\lambda = 1.1$ gave the same result. This is regarded as an indication that the sensor should survive at least 10 years in a domestic heater environment. Such an installation could easily be arranged to cut off a gas supply in the event of incomplete combustion with its attendant toxicity danger.

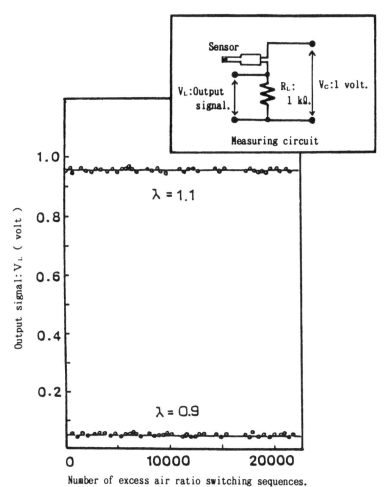

FIGURE 5.10 Repeatability of a stannic oxide combustion sensor. Fuel gas: methane; sensor and exhaust gas temperature: 600°C; switching period: 6 min.

5.3 DOMESTIC GAS HEATER MONITORING AND CONTROL

The detection of incomplete combustion in domestic heaters of various types serves as an excellent example of the use of the combustion sensor in cases where both oxygen deficiency and carbon monoxide build-up can easily occur. Various types of domestic heaters exist around the world for heating both air and water. Early forms of piped gas were usually coal gas (largely carbon monoxide and hydrogen) or town gas (a mixture of coal gas and methane-containing water gas), but, currently, natural gas (almost all methane) is predominant. Also, bottled gas (propane or isobutane) is used where piped gas is unavailable.

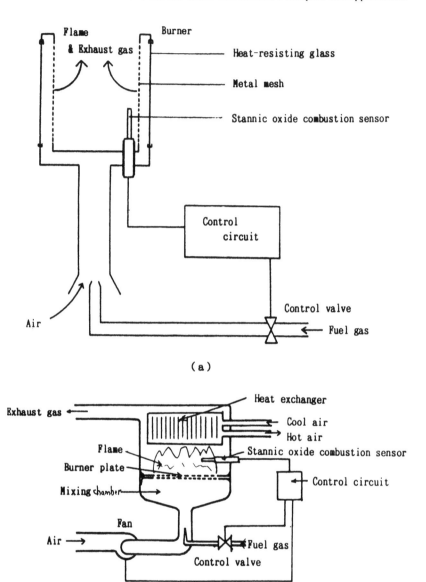

FIGURE 5.11 An incomplete combustion detector system showing sensor positioning for two types of burner.

The burning of these gases is performed in two major ways, which are shown in Figure 5.11. In diagram (a), enough air is mixed with the fuel gas to allow complete combustion inside a mesh-enclosed volume, and the sensor is situated inside the flame adjacent to the mesh. In the case of incomplete combustion taking place (perhaps due to an obstructed air inlet or an oxygen

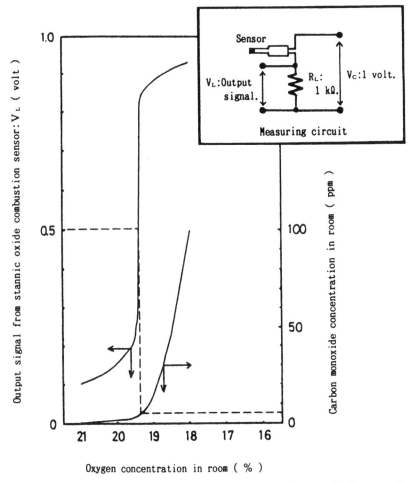

FIGURE 5.12 Change in output signal of combustion detector and increase of carbon monoxide concentration in room caused by oxygen consumption. Volume of room: 3.65 m³; combustion device: domestic gas space heater (3600 kcal/h); fuel gas: natural gas (main component: methane).

deficiency in the atmosphere), the relevant detector circuit can be designed to alarm and/or shut off the gas supply.

Diagram (b) depicts the alternative form of burner in which air is not only mixed with the fuel gas, but is freely available at the flame itself. This obviously calls for more careful positioning of the sensor, for it could be affected by the presence of this surrounding atmosphere.

In both cases, the build-up of carbon monoxide and/or the generation of an oxygen deficiency to danger levels can be prevented. (Also, it is possible to design the relevant electronics to indicate the absence of the flame altogether, for the sensor resistance will then be at its highest.)

A practical example will illustrate the efficacy of the technique. A combustion sensor was incorporated in a simple circuit as shown in Figure 5.12 (inset) and was used in conjuction with a 3600 kcal/hr gas space heater which used a burner, as in Figure 5.11(a). This was placed in a sealed 3.65 m³ room. The heater was lit, and both the oxygen deficiency and the carbon monoxide concentration in the room were monitored. The results of this test are plotted in Figure 5.12, which shows that when the oxygen concentration fell to 19.3%, the output voltage rose abruptly, indicating that the excess air ratio fell below unity of this point.

If an alarm point were set at 0.5 V for this sensor and its circuit, this would correspond to a carbon monoxide concentration of only 5 ppm (as shown by the dashed line in Figure 5.12) which is well below the TLV levels quoted earlier.

REFERENCES

1. Goto, K., Oxygen sensor based on solid electrolyte, *Denki Kagaku (J. Electrochem. Soc. Jpn.)*, 48, 361, 1980.
2. Saji, K., Takahashi, H., Kondo H., and Igarishi, I., Limiting current type oxygen sensor, Proc. 4th. Sensor Symp., Tsukuba, May 1984, 147.
3. Gibbons, E.F. *et al.*, Automotive exhaust sensor using titania ceramic, SAE Tech. Paper Series, 750224, 1975.
4. Logothetis, E.M. *et al.*, Oxygen sensors using CoO ceramics, Appl. Phys. Lett., 26, 209, 1975.
5. Ihokura, K. *et al.*, Combustion monitor sensor based on tin dioxide, *Denki Kagaku (J. Electrochem. Soc. Jpn.)*, 51, 672, 1983.
6. Ihokura, K. *et al.*, Use of tin dioxide to control a domestic gas heater, *Sensors Actuators*, 4, 607, 1983.
7. Eastwood, P.G., Claypole, T.C., Watson, J., and Coles., G.S.V., The behaviour of tin dioxide sensors in exhaust environments at low and intermediate temperatures, *Inst. Phys. Meas. Sci. Technol.*, 4, 524, 1993.
8. Takami, A. *et al.*, Effect of precious metal catalysts on TiO_2 thick film HEGO sensor with multi-layer alumina substrate, SAE Tech. Paper Series 870290, 1987.
9. Seiyama, T. *et al.*, *Kagaku-sensa (Chemical Sensors)*, Koudansya, 1982, chap. 2.
10. Yamazoe, N. *et al.*, Interactions of tin dioxide surface with O_2, H_2O and H_2, *Surf. Sci.*, 86, 335, 1979.
11. Esper, M.J. *et al.*, Titania exhaust gas sensor for automotive applications, SAE Tech. Paper Series, 790140, 1979.

CHAPTER SIX

The Domestic Gas Alarm

6.1   INTRODUCTION

The widest application of the stannic oxide gas sensor has been in the area of domestic gas alarms, notably in those countries where it is legally mandated. In this context, the detection of three basic forms of gas must be considered. Traditionally, coal gas or town gas has been extensively piped into domestic, commercial, and industrial premises. Coal gas consists largely of carbon monoxide and hydrogen, while town gas is a mixture of coal gas and water gas, and so contains methane also. However, these have been replaced in most countries by natural gas, which is largely methane. Finally, bottled gas is usually propane or isobutane.

Given these facts, it is clear that (highly toxic) carbon monoxide can build up in confined spaces, either directly from coal gas leaks or via the incomplete combustion of methane. Hence, in many applications, alarms for toxic levels of carbon monoxide are desirable in addition to methane alarms, though only the latter will be considered below in the domestic context.

6.1.1   Response to Methane

Because methane is a fairly stable gas, a stannic oxide sensor without additives is not very sensitive to it, as is shown in Figure 3.9. This is acceptable if other gases do not coexist, but it is obviously better to use a sensor which confers some selectivity to methane. A sensor developed for this purpose is the Figaro TGS 842-NG, the gas concentration characteristics for which are given in Figure 6.1.[1] Notice here the selectivity to methane compared with ethanol: this is important in domestic installations where alcohol vapor can be a cause of false alarms.

Figure 6.2 shows the dependency of this sensor on atmospheric temperature and relative humidity. Here, a reference environment of 3500 ppm of methane in air at 20°C and 65% RH gives rise to the sensor resistance R_0, while sensor resistances at other temperatures and RH values (but still in 3500 ppm of methane) are measured as R_S. Hence, resistance ratios R_S/R_0 at various temperatures can be plotted as a family of curves at several different RH values, as is shown. Notice that both low temperatures and low RH values increase the value of R_S, while high temperatures and humidities lower it. Thus, at -10°C, but maintaining the RH at 65% raises the sensor resistance to 1.2 R_0, while the combination of 40°C and 95% RH lowers it to 0.84 R_0.

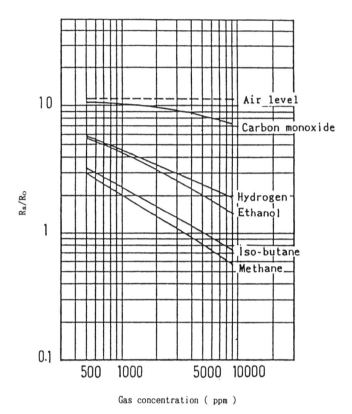

Gas concentration (ppm)

FIGURE 6.1   Gas concentration characteristics for a methane-selective sensor type TGS 842. R_o: sensor resistance in 3500 ppm methane; R_S: sensor resistance in gas.

From these examples, it is evident that safety would not be compromised by including this sensor within a simple circuit in which no temperature or humidity compensation is incorporated.

However, it does remain necessary to compensate for changes in mains voltage because these directly affect the heater voltage and hence the working temperature of the sensor. Changes in the resistance ratio can be plotted against percentage changes in the heater voltage, and this must obviously be done in a still air environment so that the effects of forced convection can be eliminated. Figure 6.3 is such a graph where the 3500 ppm methane concentration has been maintained, as have the ambient conditions of 20°C and 65% RH. The large resultant resistance ratio changes clearly establish the need for a stabilized heater voltage supply, and an appropriate circuit is given in Figure 6.4. This will be treated in detail in Section 6.3.2.

Figure 6.5 illustrates the response speed of the TGS 842; in fact, both insertion into and removal from air contaminated with 3000 ppm of methane leads to equilibrium attainment within 20 seconds, which is more than adequate for a domestic sensor.

Over a very long period, the consistency of response for the TGS 842 is illustrated in Figure 6.6. Here, it will be seen that there is no discernable

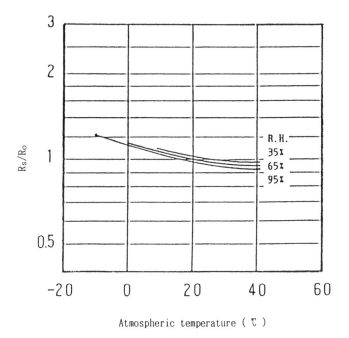

FIGURE 6.2   The dependency on atmospheric temperature and humidity of TGS 842 characteristics. R_o: sensor resistance in 3500 ppm methane at 20°C and 65% RH; R_S: sensor resistance in 3500 ppm methane at various atmospheric temperatures and humidities.

degradation in performance for the period of the trial at the time of writing, which is four years of continuous energization.

6.2   REQUIREMENTS FOR A DOMESTIC GAS DETECTOR

Specifications for a domestic gas detector may be written after due consideration of various basic requirements which have been promulgated in several countries, some of which are now mandatory in those countries.

6.2.1   Physical Requirements for a Domestic Gas Detector

The well-known battery-operated ionization-type smoke detector has been installed in domestic and commercial premises in numerous countries for some considerable time. However, an analogous detector for piped or bottled gas is less common (though it is now a legal requirement in Japan). Mains-operated stannic oxide sensor instruments are appropriate to this latter application for several reasons, which will be presented below, but at the time of writing, there is little possibility of designing a battery-operated version (except for intermittent usage) because of the necessary heater current drain. This may not be disadvantageous, however, because it eliminates the necessity for low-battery alarm circuitry and also makes possible the inclusion of other current-consuming components, such as relays for remote indication or actuation.

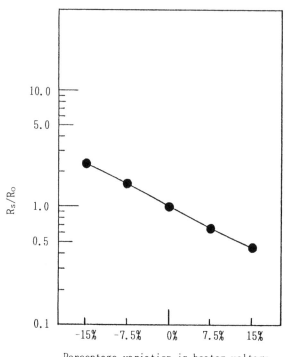

Percentage variation in heater voltage

FIGURE 6.3   The effect of heater voltage variations on the TGS 842 sensor. R_o: sensor resistance in 3500 ppm methane at standard heater voltage; R_S: sensor resistance in 3500 ppm methane at other heater voltages; ambient temperature: 20°C; relative humidity: 65%.

However, the domestic situation is one in which maintenance can be notoriously ineffective! For example, the gas detector can be left in an inoperative state for extensive periods; it is quite unrealistic to expect the average householder to obtain and use calibration equipment (such as a test gas container) at appropriate intervals. Hence, a domestic gas detector must be long-lived and robust, and it must maintain its alarm accuracy over a very long time (which in Japan has been defined as a minimum of five years). Fortunately, the stannic oxide sensor can fulfill such requirements given appropriate circuitry and housing. Apart from its well-established longevity, it can be set to alarm at a gas concentration well below danger level, so that if it eventually drifts somewhat, it will still alarm well below this danger level, and safety will not be compromised.

6.2.2 Specific Requirements

Table 6.1 summarizes the specific requirements for domestic gas alarms published by the appropriate bodies in three countries and is valid at the time of writing. The following notes show how a current domestic sensor — the methane-sensitive TGS 842 — meets these requirements.

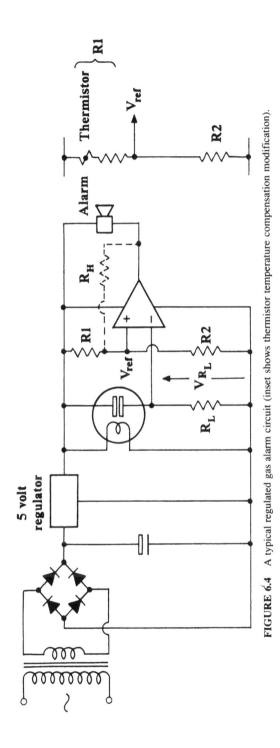

FIGURE 6.4 A typical regulated gas alarm circuit (inset shows thermistor temperature compensation modification).

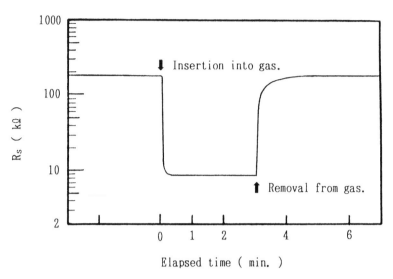

Elapsed time (min.)

FIGURE 6.5   The response speed to 3000 ppm methane of TGS 842.

The longevity of the stannic oxide sensor has already been demonstrated in Figures 2.29 and 2.30, and these are now augmented by Figure 6.6, which refers specifically to the TGS 842.

The limits of gas concentration within which an alarm must operate should first be defined under specified conditions of temperature and relative humidity and must then be allowed to depart from this ideal because of the inevitable fluctuations in these parameters and in the power supply voltage. Allowed limits of sensor degradation with time must also be part of such a specification. The final "band of alarm operation" must have an upper limit well below danger level (either explosive or toxic, depending upon the particular gas involved) but a lower limit high enough to avoid most false alarms resulting from activities such as painting, wine-making or smoking.

It is possible to design gas-detection electronic circuits which will compensate for power supply variations, temperature, and RH changes (in increasing order of difficulty), but these inevitably lead to increased costs, and eventually a point is reached where the resulting instrument ceases to be commercially viable. This is especially true if an inbuilt storage battery back-up is incorporated along with a mains-failure alarm.

These factors will now be considered separately.

Firstly, Table 6.1 indicates that an instrument which alarms between 2500 and 10,000 ppm (0.25 to 1%) is considered satisfactory, and it is usual to truncate this to about 3500 to 4000 ppm in actual instruments. This leaves considerable latitude for variations due to changes in temperature and RH, but when setting an instrument to operate in such a closely defined band, the long-term transient must also be taken into account as explained in Section 2.6.2. If this is not done, and an instrument is preset before the sensor has stabilized, then obviously the alarm band will shift during actual usage.

TABLE 6.1 Extracts from Various Domestic Gas Alarm Standards*

	Inspection standards for town gas leakage alarm; (Japan Gas Appliances Inspection Association) Japan	Specification for the electrical apparatus for the detection of combustible gases in domestic premises; (British standard BS 7348) U.K.	Residential gas detectors. UL 1484 standard for safety; (Underwriters Laboratories Inc.) U.S.A.
Alarm concentration in standard atmosphere	1/200 ~ 1/4 of LEL. Methane: 0.1 ~ 1.25%. (1000 ~ 12,500 ppm)	5 ~ 20% of LEL. Methane: 0.25 ~ 1.0%. (2500 ~ 10,000 ppm)	Less than 25% of LEL; natural gas: 0.95% (the LEL of N.G. specified as 3.8% in this standard) Fluctuating upper limit: $U = (K+I) / 2$; K: 25% of LEL, I: Initial alrm. conc.
Fluctuation of alarm conc. by atmospheric temp. and humidity	−10°C ~ 50°C; alarm as above. 35 ~ 40°C, >85% RH; alarm as above. 50°C; clear of false alarms by miscellaneous gases	Not specified precisely. The above is to be maintained from 15°C, 35% RH to 25°C, 60% RH.	Alarm as above; under all the following conditions: 0°C, 30% RH. ~ 49°C, 50% RH; 30°C, 0% ~ 90% RH; and at 42°C, 95% RH.
Change in alarm conc. by supply voltage fluctuations	Fluctuation of ±10%; alarm as above	Fluctuation of +10% and −15% of 240 V; alarm as above	Fluctuation of +10% and −15%; as above
Long-term stability	After one month or more storage with and without energizing; alarm as above and clear of false alarms by miscellaneous gases; Durability for over 5 years	After 3 months being subjected to 5% and 20% of LEL once a week; alarm as above	Alarm as above during three months observation being subjected to temperature change, gas flow velocity change, etc.
Response speed to alarm	At 1/4 LEL (methane 1.25%); alarm within 20 sec	At 25% of LEL (methane 1.25%); alarm within 30 sec	Not specified
False alarm by miscellaneous gases	For ethanol of 0.1% or less at 38 ~ 40°C, >85% RH; no alarm.	Not specified	Not specified
Explosion proofing of sensor housing	Proofed with double stainless steel gauze as specified or with more strict flame arrester	Same as Japanese specification; no ignition for 10 times sparking in a mixture of 28 ± 1% hydrogen in air	No ignition for 10 min energizing with 110% supply voltage in 8.3% natural gas

LEL: lower explosion limit.

*For comparative purposes only; original documents must be consulted for design compliance.

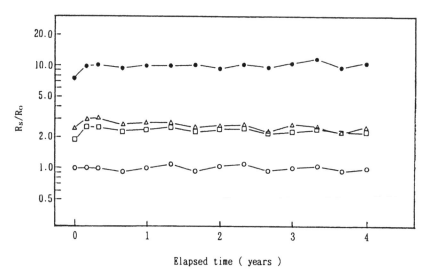

Elapsed time (years)

FIGURE 6.6 Long-term operation of the TGS 842. Gas: (O): methane; (□): ethanol; (Δ): hydrogen; (●): air. Gas concentration: 3000 ppm. R_o: sensor resistance at first measurement for methane; R_s: sensor resistance at each measurement for all gases.

Secondly, compensation for temperature and humidity changes must be considered. In many countries, room temperature can fall below -10°C, especially briefly at night or in the absence of the occupants. Conversely, on a hot day, the temperature may exceed 30°C, and this may be exacerbated by cooking procedures, particularly when the detector is mounted near a ceiling, as it should be since methane is less dense than air and so collects at the higher levels.

Temperature compensation is relatively easily achieved using thermistor techniques (as will be shown), but RH compensation is much more difficult, since there is no humidity sensor available which is sufficiently fast, reliable, and cheap. One solution has been to place a permanently wet wick adjacent to the gas sensor, so establishing a very high localized RH at all times. This, however, implies periodic topping up and changing of a reservoir, which rules out the method for domestic installations.

Fortunately, in many geographical areas — but not all — temperature and humidity tend to rise or fall together, so that an "over-compensating" thermistor circuit may be used satisfactorily. Even more fortunately, sensors such as the TGS 842 can be fabricated to have very low dependencies on temperature and RH (as shown in Figure 6.2) and so may need no compensation at all. This is again, because a well-aged sensor of this type can be incorporated in a circuit which will alarm over a narrow range, well within the relevant allowed band.

Thirdly, all countries experience power supply voltage fluctuations, and these usually occur within published limits. The most serious effect of this would be to modify the sensor temperature via concomitant heater voltage variations, which can have a profound effect on its performance, as has been shown above. Hence, heater voltage regulation is mandatory for most applications.

Fourthly, as has been mentioned, modern stannic oxide sensors have very long lives and drift very little, as shown by Figure 6.6 for the TGS 842. Furthermore, the response speed is not a factor for a domestic installation, as has also been pointed out.

Fifthly, when properly installed near a ceiling for methane detection, a sensor is likely to encounter other gases such as alcohol vapor, and this could lead to occasional false alarms which could result in a real alarm being ignored by the occupants. To avoid this situation, a reasonably high degree of selectivity is desirable, and this is exhibited by the TGS 842 as shown in Figure 6.1.

Finally, the housing of the sensor and its electronics must be "explosion-proof", which is to say that a concentration of gas well above the lower explosive limit (LEL) level, and at least four times the alarm level, should not be ignited by the instrument. All modern commercial stannic oxide sensor housings fulfill this requirement, and it is not difficult to design appropriate housing for the electronics.

6.3   DOMESTIC GAS ALARM DESIGN

As has been pointed out already, the mandatory requirements of various countries must be taken into account when domestic gas alarms are at the design stage. However, this was not always so, and in the early days of such instruments, virtually no rules nor even guidelines were available. It is therefore instructive to include brief notes on the earliest products and to highlight some of their many disadvantages.

6.3.1   Early Gas Alarm Circuits

Since 1968, more than 50 million domestic gas alarms of various designs have been manufactured. Some simply operate an internal audible alarm, but others incorporate some form of relay for the actuation of an external alarm, gas valve, or ventilation fan. More complex versions have multipoint sensors with alarms at a central station, others operate via telephone lines, and some even incorporate voice synthesizers which urge the hearer to ventilate the room or depart forthwith! However, all rely ultimately upon the reliability and longevity of the sensor and its immediate circuit.

One of the very earliest circuits is shown in Figure 6.7, and this was originally intended for the Japanese AC mains supply. It incorporated a simple step-down transformer to energize the (1 V) heater, but applied the full mains across the sensor itself to drive a buzzer. Such alarms were susceptible to mains voltage variations, being totally unregulated, and were notorious for alarming in the presence of gases such as alcohol vapor, carbon monoxide (from smoking), some paint and insecticide fumes and, of course, water vapor. There were also some reports of the sensor becoming red hot by self-heating (especially when the circuit was modified for higher-voltage European mains, and if low-resistance buzzers were used) so that they never switched off!

Figure 6.8 has been included as an example of the performance of a gas alarm based upon such a primitive circuit, but using the modern Figaro TGS

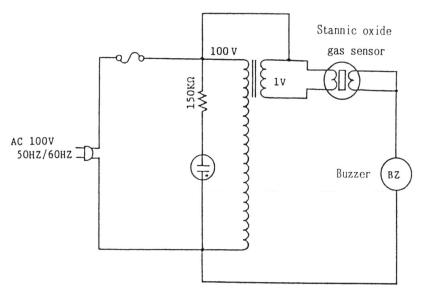

FIGURE 6.7   An early domestic gas alarm circuit. Gas sensor: Figaro Engineering Inc. TGS 109.

842 sensor, and it will be seen that the alarm point changed from 3500 ppm to over 14000 ppm for a 15% drop in mains voltage! This clearly demonstrates the necessity for more sophisticated circuits such as that of Figure 6.4.

6.3.2 Modern Electronic Circuits

In Figure 6.4, the AC from a transformer is rectified and capacitively smoothed as shown, and the resultant DC fed to a 5-V regulator which supplies the entire circuit. Such regulators are both reliable and inexpensive and very easily obtainable because 5 V is the standard logic circuit supply voltage.

Such a regulator needs to drop a specified minimum voltage $V_{R(min)}$ from input to output in order to operate satisfactorily, and this defines the lowest voltage which may be applied from the rectifier, which in turn defines the minimum acceptable r.m.s. voltage provided by the transformer secondary. This, of course, occurs when the mains voltage is at its allowed minimum. If the minimum rectified and smoothed voltage to the regulator is $V_{in(min)}$, then

$$V_{in(min)} = V_{R(min)} + 5 \qquad (6.1)$$

The maximum power dissipated by the regulator will occur when its input voltage is maximal and when the total current to the circuit is also maximal, which is when the alarm is operating:

$$P_{R(max)} = V_{R(max)} \cdot I_{(max)} \qquad (6.2)$$

It is under this condition that maximum heat will be generated, so that the cooling fins and thermal bonding of the regulator must be adequate. This is not

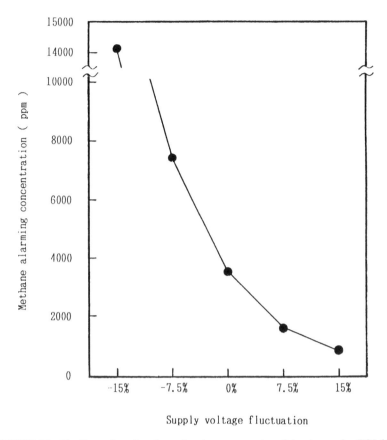

FIGURE 6.8   The fluctuation of methane alarming concentration of the alarm using TGS 842 without compensation for supply voltage fluctuation.

normally a problem, however, because even under these worst-case conditions, the power dissipated is not large.

The regulated 5-V supply drives both the heater and the subsequent electronics. Hence, the comparator (or operational amplifier connected as a comparator) must be a single-supply type intended to work at this voltage. Then, the circuit must be designed so that all voltages are positive, otherwise a negative supply would be needed. This has been done in Figure 6.4, as follows.

A reference voltage V_{ref} derived from the potentiometer R1 and R2 is applied to the noninverting input of the comparator, and if R1 = R2, then $V_{ref} = 2.5$ V. Then the sensor load resistance R_L is chosen so that at the nominal alarm gas concentration (such as 3500 ppm), it will also drop 2.5 V, and this voltage is applied to the inverting input of the comparator. This means that as V_{R_L} slightly exceeds V_{ref}, the output switches rapidly from nearly 5 V to nearly 0 V, and the alarm operates.

Note that there could be comparator indecision at this point if the gas concentration fluctuated somewhat, resulting in the alarm turning on and off spasmodically. It is therefore advisable to insert a hysteresis resistor R_H as shown dashed in Figure 6.4[2]

This appears in parallel with R1 when the output is high, and in parallel with R2 when it is low, so that V_{ref} is a little lower when the alarm is on than when it is off. Hence, the gas concentration must fall just below this lower alarm point before the alarm turns off. The width of this "window of operation" depends on the value of R_H compared with R1 or R2, and typically it is about an order of magnitude greater.

The two values of V_{ref} will be

$$V_{ref\,(High)} = \frac{5\,R2}{\dfrac{R1 \cdot R_H}{R1 + R_H} + R2} \tag{6.3}$$

and

$$V_{ref\,(Low)} = \frac{5\,\dfrac{R2 \cdot R_H}{R2 + R_H}}{R1 + \dfrac{R2 \cdot R_H}{R2 + R_H}} \tag{6.4}$$

By this means, the alarm could be made to turn on at $V_{R_L} = 2.6$ V and turn off at $V_{R_L} = 2.4$ V, for example. The "window" of 0.2 V would then correspond to a small gas concentration difference which would depend on the sensitivity of the sensor.

The choice of R_L is also dependent upon this sensitivity, as an example will show. If the air level resistance of a sensor were 120 kΩ and the resistance ratio R_s/R_a at 3500 ppm of methane were 0.05, then:

$$R_S = 120 \times 0.05 = 6 \text{ k}\Omega \text{ at 3500 ppm}$$

Hence, to produce a voltage drop of 2.5 V using a 5-V supply, R_L must also be 6 kΩ. To take account of tolerance spreads in the sensor, it would be reasonable to use either a 10 kΩ preset variable resistor (or perhaps a 5-kΩ variable in series with a 4.9-kΩ fixed resistor). This would allow for the alarm calibration of each unit.

If it is desired to incorporate temperature compensation into the circuit of Figure 6.4, this may be accomplished by including a thermistor in series with R1 as shown in the inset diagram. Here, if 10 kΩ is suggested as a suitable value for the potentiometer resistors, then R1 may be made up of a 6.8 kΩ fixed resistor in series with a thermistor which exhibits a resistance of about 3.3 kΩ at 20°C. The efficacy of the circuit may then be evaluated empirically at a series of ambient temperatures, keeping in mind that for full consistency of operation, the sensor should be energized over a period prior to this evaluation (which may be only one week for the TGS 842, but up to four weeks for some other sensors).

6.3.3 Operational Design Factors for Domestic Gas Alarms

It has been noted in the previous section that thermistor temperature compensation can be applied to an appropriate drive circuit such as that in Figure 6.4. In fact, it will be found that this technique will also compensate, to some considerable extent, for changes in RH, too, and this can also be checked empirically using appropriate test apparatus.

However, considerable care must be taken when deciding each temperature/RH combination to use for test purposes. This is because ambient temperature and RH are inextricably linked, and the way they vary is dependent upon the climate of the area in question and the time of year. However, if a closed test chamber with an atmosphere at 20°C/65% RH were heated up to (for example) 35°C, the RH would fall to 27%, which is unlikely to be paralleled meteorologically.

Had the absolute humidity been measured (that is, the amount of water per unit volume of air), this would have remained constant as the chamber was heated. Hence, there is a case for characterizing gas sensors in terms of absolute rather than relative humidity.

Looked at the other way round, an atmosphere of 35°C/65% RH contains 2.4 times more water than an atmosphere at 20°C/65% RH.

As can now be seen, temperature and humidity compensation, though apparently simple from a circuit aspect, is not only difficult to calibrate, but the calibration needed is very dependent upon the temperature/RH relationship to be expected in the area of usage. Fortunately, because the eventual instrument is to be used inside domestic premises, it can reasonably be calibrated in terms of the "comfort zone", and set to alarm at a level low enough to present no hazard even at the conceivable extremes of the temperature/RH combination spectrum.

Normally, the instrument will be permanently energized, but at initial switch-on, the alarm will sound for the duration of the short-term transient. This is actually a useful indication that the instrument is working properly, but in cases where it is desirable to mute the alarm for this period, a simple timing circuit can be designed to temporarily short-circuit R1. (An FET-input 555 timer will conveniently do this.)

The placement of the instrument is important when considering possible false alarms resulting from miscellaneous gases. For example, the instrument should be placed at a high point to detect methane (or other light gases such as hydrogen), but it may then respond to alcohol vapor, which also collects near ceilings. For a low placement to detect heavy bottled gases such as propane, however, problems may arise with organic solvents as used in insecticides or perfumes, or as spraying agents in aerosol bottles. In both cases, the selectivity characteristics of sensors such as the Figaro TGS 842 become very important, for it has been seen that the response of a stannic-oxide based sensor to such "nuisance" gases cannot be completely eliminated.

One helpful approach is to delay alarm operation until some 20 to 30 seconds after the alarm level has been reached. This will often be long enough for any transient interference by sprayed insecticide, perfume, or aerosol to dissipate. (Again, a 555 timer can be used in such a circuit.)

6.3.4 Some Manufacturing Considerations.

Because a domestic alarm is almost never recalibrated after leaving a production line, care must be taken to ensure the initially set alarm point is both correct and stable. To this end, two important points must be addressed.

Firstly, calibration must take place only after the sensor has been operated (or "aged") long enough stabilize sufficiently, which may take between a week and a month, depending upon the sensor type. (This has been covered in Section 2.6.2 on the long-term transient.)

Secondly, the sensor must be operated in an atmosphere contaminated to just below the alarm point during the aging process.

It is possible — and usual — for batches of sensors to be aged, then incorporated into their instruments prior to calibration. By this means, the individual sensitivities of the sensors are each taken into account. The (cheaper) alternative would be to preset all the electronic circuits and simply insert each aged sensor as a component, but this inevitably leads to a wide spread in alarm points. This is further exacerbated by the fact that the heater voltage produced by the stabilizer in each instrument is also subject to a tolerance (often 5 ± O.2 V) and so may differ somewhat from the heater voltage applied during ageing. In this case, the sensor will eventually reach a slightly different surface

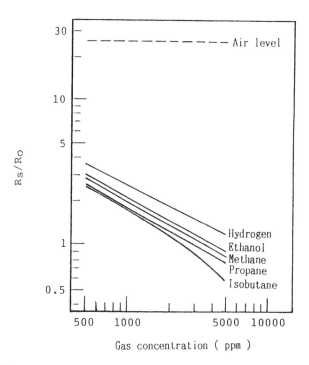

FIGURE 6.9   Gas concentration characteristics of a sensor for bottled gas (TGS 109). R_o: sensor resistance in 3500 ppm methane; R_s: sensor resistance in gas.

equilibrium condition *in situ* than it reached during the aging process, so leading to a change in sensitivity.

Clearly, the method chosen for aging and calibrating is the result of a compromise between cost and alarm point accuracy, repeatability, and invariance over long periods.

6.4 DOMESTIC GAS ALARMS FOR BOTTLED GAS AND TOWN GAS

The design of alarms for bottled or town gas is not significantly different from that for methane. However, in the former case, a sensor with high sensitivity to propane and butane is required, and in the latter case, to hydrogen, which is a major component in the rather complex mixture typical of town gas.

Bottled gas is denser than air, so the alarm should be placed near floor level, where it is unlikely to meet either methane or hydrogen, so that extreme selectivity is not mandatory. Figure 6.9 shows characteristics of the Figaro TGS 109 sensor and indicates that it is entirely appropriate for this application. Table 6.2 is part of the current Japanese certification standards for bottled gas alarms. Figure 6.10 shows the characteristics of a similar sensor modified so that its sensitivity to hydrogen is improved, especially by comparison with that to ethanol. This sensor is therefore appropriate for use in town gas alarms, the circuitry for which can be identical to that for the bottled gas versions.

TABLE 6.2   Extract from Japanese Certification Standards for Liquid Petroleum Gas Leakage Alarms (High Pressure Gas Safety Institute of Japan)

Parameter	Standard
Alarm concentration in standard atmosphere	1/100 ~ 1/4 of LEL
	isobutane: 0.05 ~ 0.30%
	(500 ~ 3000 ppm)
Fluctuation of alarm concentration by	−10°C; 0.05 ~ 0.35%
atmospheric temperature and humidity	(500 ~ 3500 ppm)
	>40°C, >85% RH; 0.05 ~ 0.30%
	(500 ~ 3000 ppm)
Change in alarm concentration by supply	Fluctuation of ±10°; alarm as above
voltage fluctuations	
Long-term stability	After two months or more of storage without
	energizing: alarm as above
Response speed to alarm	At 0.45% of isobutane: alarm within 30 sec
Explosion proofing of sensor housing	100 times discharge spark in the sensor
	housing should not ignite 2.5 ~ 3.5% of
	isobutane in surrounding atmosphere
False alarm by miscellaneous gases	Not specified

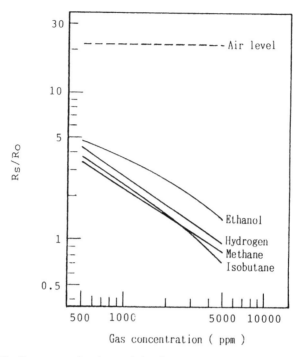

FIGURE 6.10   Gas concentration characteristics of a sensor for town gas (TGS 109M). R_o: sensor resistance in 3500 ppm methane; R_s: sensor resistance in gas.

REFERENCES

1. TGS 842 Catalog Sheet, Figaro Engineering Inc.
2. Watson, J., *Analog and Switching Circuit Design*, Wiley Interscience, New York, 1989, 394, 589.

CHAPTER SEVEN

Industrial Gas Sensors and Instruments

7.1 INDUSTRIAL GAS SENSORS

Historically, stannic oxide gas sensors were developed with the domestic market in mind and so were designed basically for inexpensive mass production. More recently, as their *modus operandi* became better understood, it became possible to manufacture them to closer tolerances and within much more robust,[1] and also special-purpose, encapsulations. Such devices will now be described and may be compared with the less expensive versions already introduced in Chapter 1.

7.1.1 Industrial Sensor Encapsulations

For use in environments involving high ambient temperatures, and/or chemically active components in the atmosphere, encapsulations constructed from metal and alumina alone have been developed, with no plastic content, as illustrated in Figure 7.1. This form of structure is very robust, and its shock-resistant properties are fully specified.

Conversely, for the monitoring of carbon monoxide, a less robust, but special-purpose encapsulation is appropriate, as shown in Figure 7.2. Here, a plastic "body" is adequate, but it incorporates an activated charcoal filter to remove any alcohol (ethanol) vapor which may be present.[1,2] This is because sensor material which is sensitive to carbon monoxide is also sensitive to ethanol vapor. Normally, the presence of ethanol vapor is transitory, so that any absorbed by the charcoal will later be driven off by the warm sensor. Hence, the filter is effectively self-regenerating. In the continuous presence of ethanol vapor, however, the filter will become saturated, so that the sensor is obviously not appropriate for use in these circumstances.

7.1.2 Structure of the Active Element

Examples of both directly and indirectly heated sensor elements have been described in Chapter 1, and both are sufficiently robust to incorporate into metal/alumina industrial encapsulations. However, there are special cases where the type of sensor structure may be defined by other considerations. For example (and as will be explained later), the carbon monoxide sensitive Figaro

159

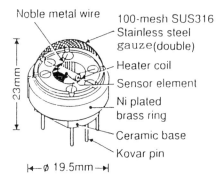

FIGURE 7.1 The construction of a heat and chemical resistant sensor.

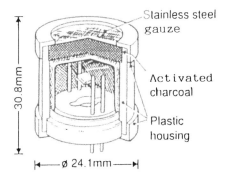

FIGURE 7.2 The construction of a sensor for low level carbon monoxide.

TGS 203 is cyclically heated to two different temperatures, and it has been found that the directly heated structure is more tolerant to this form of thermal shock than is the indirectly heated equivalent.

Another example is a small low-power version developed for the sensing of sulfides and mercaptans (and often used in halitosis checking instruments!). This is shown in Figure 7.3[1,3] and has a very low thermal capacity so that it loses little heat by conduction through the structure, which results in a power consumption of only about 14 mW. It is manufactured by forming an alumina insulating layer on a single Kanthal heater wire, followed by the deposition of a thin film of stannic oxide ceramic. This element is bonded at three points with gold paste onto a metal/alumina substrate, as shown.

7.2 A LOW-LEVEL CARBON MONOXIDE MONITOR

To illustrate the development of a sensor and its circuitry for a specific requirement, the monitoring of carbon monoxide provides an excellent example, for it is not only a very toxic, but a very common, gas.

7.2.1 The Toxicity of Carbon Monoxide

Figure 7.4 shows the effects on the human body of various levels of carbon monoxide in the atmosphere.[4] If carbon monoxide is present, and is respired,

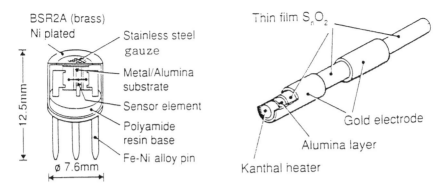

Sensor housing

Sensor element structure.

FIGURE 7.3 The construction of a low-power sensor.

it is taken up by blood in preference to oxygen, forming carboxyhemoglobin. This causes progressively more severe symptoms as its concentration increases, culminating in death; but often the early symptoms are not recognized by the sufferer, so rendering the danger even greater.

Below 10% carboxyhemoglobin, the physiological effect is negligible, and this is illustrated by the curve "I" in Figure 7.4. This corresponds to about 60 ppm of carbon monoxide in the atmosphere, which is why the tolerance limit value (TLV) level is quoted as 50 ppm in Europe and Japan (but is somewhat lower at 35 ppm in the U.S.) at the time of writing.

The time of exposure is also a factor. For example, at 400 ppm, some 15 minutes must elapse before the 10% carboxyhemoglobin level in the blood of an average adult is reached. Hence, a monitor which alarms within 15 minutes at 400 ppm can trigger personnel departure long before any danger is incurred. Considerations such as this have been taken into account when formulating the proposed standards for carbon monoxide monitors shown in Table 7.1.

7.2.2 A Low-Level Carbon Monoxide Sensor

From the previous section, it is clear that a sensor which can reliably detect 10 to 200 ppm of carbon monoxide in the atmosphere is required. This can reasonably be based on the sensor material described in Section 4.2, and when encapsulated along with an activated charcoal filter as shown in Fig 7.2, it has been commercialized as the Figaro TGS 203.

Fig 4.8 shows that for this sensor to achieve maximum selectivity to carbon monoxide, the operating temperature should be below 100°C. Unfortunately, at such a low temperature, not only is the response slow, but the sensor can easily accumulate miscellaneous gases which interfere with its operation. It is therefore heated sequentially to a high purging temperature (260 to 300°C), followed by a low operating temperature (about 80 to 100°C). At the end of each operating period, the conductivity of the active material is rapidly measured, and the cycle then repeats. When working under these conditions, and in its

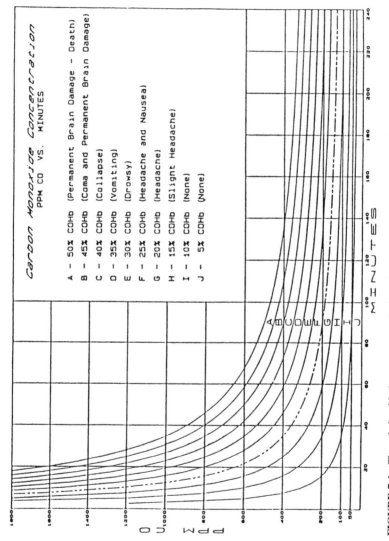

FIGURE 7.4 The relationship between atmospheric carbon monoxide concentration and exposure time leading to a carboxylhemoglobin (COHb) concentration in blood.

TABLE 7.1 Extracts from Domestic Carbon Monoxide Gas Alarm Standards*

	Inspection standards for incomplete combustion alarm (CO detecting type). (Japan Gas Appliances Inspection Association) Japan	Proposed first edition of the standard for single and multiple station carbon monoxide detectors. UL 2034. (Underwriters Laboratories Inc.) U.S.A.
Alarm concentration in standard atmosphere.	50 ppm CO: no alarm 200 ppm CO: alarm within 15 min 250 ppm CO: alarm within 5 min	100 ppm CO: alarm within 120 min 200 ppm CO: alarm within 45 min 400 ppm CO: alarm within 15 min 100 ± 5 ppm CO: no alarm for first 5 min $15 + 3, -5$ ppm CO: no alarm for 24 hours
Fluctuation of alarm concentration by atmospheric temperature and humidity.	$\leq 0°C$: alarm as above $\geq 50°C$, $35 \sim 45\%$ RH: alarm as above $45 \sim 50°C$, $\geq 60\%$ RH: alarm as above	As above under all the following conditions: 49°C, $30 \sim 50\%$ RH, O_2: $20 = 1\%$ 0°C, $15 \pm 5\%$ RH, O_2: $20 = 1\%$ $52 \pm 3°C$, $95 \pm 4\%$ RH
Change in alarm concentration by supply voltage fluctuations	Fluctuation of $\pm 10\%$: alarm as above and clear of false alarm by ethanol below	Fluctuation of $+10\%$ and -15%: alarm as above
Long-term stability or durability	After 10 days storage with power in 35°C, 60% RH. exposing to 500 ppm H_2 for 30 min twice a day: no alarm for 25 ppm CO, alarm within 15 min for 200 ppm CO and within 5 min for 550 ppm and clear of false alarm below.	Alarm as above after exposing to 0.1% moist H_2S or 1% moist CO_2-SO_2 for 10 days while de-energized
False alarm by miscellaneous gases	Coexisting 30 ppm NO: alarm as above 500 ppm H_2: no alarm for 5 min 1000 ppm ethanol at 40°C, 85% RH: no alarm for 15 min	Not specified
Explosion proofing of sensor housing	Proofed with double stainless steel gauze as specified or with more strict flame arrester.	Not specified

*For comparative purposes only; original documents must be consulted for design compliance.

standard encapsulation, the gas concentration characteristics for the TGS 203
appear as shown in Fig 7.5

7.2.3 Drive Requirements and Circuit Design for a Carbon Monoxide Monitor

Clearly, two different heater voltages must be applied to the TGS 203 to
achieve the necessary high purge and low operating temperatures, and each of
these must be applied long enough for the the two temperatures to stabilize.
Figure 7.6(a) shows that if a higher heater voltage $V_{H(Hi)} = 0.8$ V is applied for
60 seconds, the sensor temperature stabilizes at about 300°C; and if this is
followed by a lower heater voltage $V_{H(Lo)} = 0.3$ V for about 90 seconds, then
the sensor will cool and stabilize at about 80°C.

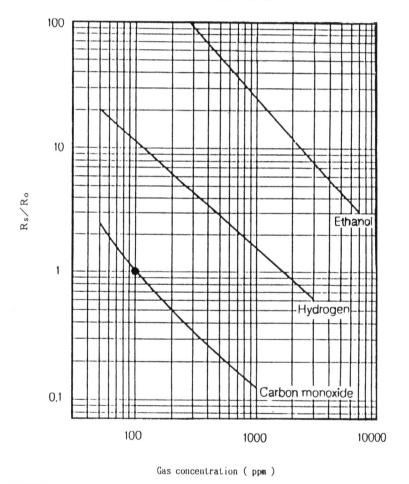

Gas concentration (ppm)

FIGURE 7.5 Gas concentration characteristics for the TGS 203. R_o: sensor resistance in 100
ppm carbon monoxide; R_s: sensor resistance in gas.

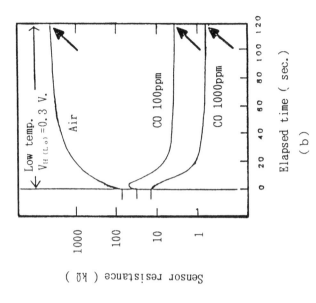

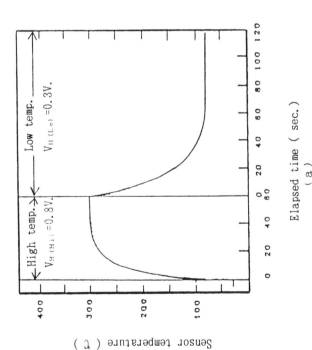

FIGURE 7.6 Sensor temperature and resistance at two heater voltage levels. Sensor: Figaro gas sensor TGS 203 carbon monoxide sensor; heater voltages: 0.8 V and 0.3 V; gas: carbon monoxide. (From Murakami, N., Dig. 5th Chem. Sens. Symp., Electrochem. Soc. of Japan, Tokyo, Sept. 1986, 53.)

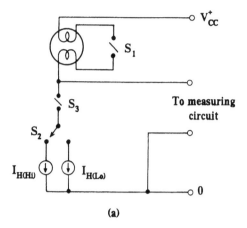

(a)

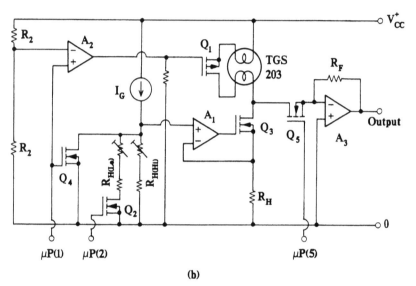

(b)

FIGURE 7.7   Basic system (a) and practical implementation (b) of a drive and measuring system for a TGS 203 Carbon Monoxide Sensor.

Figure 7.6(b) shows that this operating time of 90 seconds is adequate for the sensor resistance to stabilize in both clean air and up to 1000 ppm of carbon monoxide. It is therefore at the end of this period that a rapid measurement of R_S or G_S is taken, after which the entire 150 seconds cycle repeats. This means that the display or readout of an instrument incorporating this sensor will be updated every 2.5 minutes, which is sufficiently frequent having regard to the speed of carbon monoxide build-up in almost all situations normally encountered.

There are numerous ways of designing an electronic system to energize the two heaters for the requisite periods, then to measure the resistance or conductance between them. Basically, the heaters may be energized using one of three methods, as follows:

1. Two voltage sources may be used, each switchable between 0.3 and 0.8 V.
2. A single voltage source may be used, switchable between O.6 and 1.6 V, and with the heaters connected in series.
3. Either one or two current sources may be used to give appropriate voltage drops across the heaters, in which case their values will be approximately $I_{H(Hi)} = 370$ mA and $I_{H(Lo)} = 134$ mA.

Finally, appropriate timing circuitry must be provided, as must a method for measuring the resistance or conductance through the active material at the end of the cycle. (This involves isolating the heaters very briefly and measuring the resistance or conductance between them.)

As an example of a specific drive circuit, consider a dual current source system[5] leading to the block diagram of Figure 7.7(a). Here, the sequence of switch operation would be as follows:

Period	S1	S2	S3
Purging (60 s)	On	$I_{H(Hi)}$	On
Operating (90 s)	On	$I_{H(Lo)}$	On
Measuring (1 s)	Off	—	Off

The switches can be small-power MOSFETs and the gate drives can reasonably be derived from a microprocessor, which is then also available to perform various logistic tasks within a carbon monoxide monitoring instrument.

Figure 7.7(b) shows one practical implementation of a drive system as an example. Here, a low-drift current generator I_G feeds either the resistance $R_{H(Hi)}$ or both $R_{H(Hi)}$ and $R_{H(Lo)}$ in parallel, depending upon whether the switching MOSFET Q_2 is off or on, as controlled by microprocessor signal $\mu P(2)$. This produces either a high or low voltage drop which is applied to the voltage-to-current converter consisting of operational amplifier A_1, resistor R_H, and MOSFET Q_3. Hence, either $I_{H(Hi)}$ or $I_{H(Lo)}$ passes through both heaters held in series by (P-channel) MOSFET Q_1.

At the termination of the purging and operating periods, microprocessor signal $\mu P(1)$ briefly switches on Q_4, and this both reduces the heater current to zero and turns off Q_1 via A_2, so isolating the heaters. Microprocessor signal $\mu P(5)$ then turns on Q_5 within this brief period to sample the current through the sensor material between the heaters, and this is measured by the transresistance amplifier A_3. The final result is an output voltage proportional to the sensor conductance according to Equation 1.6, and this, in turn, is closely proportional to gas concentration at low concentration levels.

The microprocessor can be programmed to accomplish the timing sequence, and it can also operate an indicating system (such as a bank of LEDs) to inform the user how far into the 2.5 minute cycle the system has progressed. Readout is most conveniently achieved via a commercial digital panel display which will automatically hold each reading until again updated; and the microprocessor can also be programmed to implement appropriate alarm signals.

Over the range 0 to 100 ppm of carbon monoxide, the linearity of the output voltage as a function of the gas concentration has been found to be very good, so that look-up table linearization is normally unnecessary (though the micro-processor plus a ROM can easily accommodate this). Also, compensation for ambient temperature (and to some extent RH) changes can be minimized by using an appropriate thermistor circuit in place of the simple feedback resistor R_F.

REFERENCES

1. Figaro Engineering Inc. Product Catalog.
2. Murakami, N., The selective detection of carbon monoxide with a stannic oxide gas sensor using a sensor temperature cycle, Dig. 5th. Chem. Sens. Symp., Electrochem. Soc. of Japan, Tokyo, September 1986, 53.
3. Kohda, H. et al., New shape SnO_2 thin film gas sensor, Abstracts I, 1989 Int. Chem. Congr. Pacific Basin Soc. 02 ANYL 206, Honolulu, December 1989.
4. UL 2034, Proposed first edition of standard for single and multiple station carbon monoxide detectors, Underwriters Lab. Inc., Sect. 34, January 1991, 46.
5. Watson, J. and Davies, G., A low-level carbon monoxide monitor, *Sensors Actuators,* B2, 219, 1990.

CHAPTER EIGHT

Future Developments

8.1 ADVANTAGES AND DISADVANTAGES OF THE METAL OXIDE GAS SENSOR

Stannic oxide sensors — like all metal oxide sensors — have both significant advantages and disadvantages compared with other forms of gas sensor. Many of these have become apparent over several decades of practical application and have led to new and improved products via commercial feedback. The most important topics in this context are presented below, followed by some extrapolations for future developments.

8.1.1 Advantages of the Stannic Oxide Gas Sensor

Insofar as the domestic gas sensor market is concerned, the prime advantages of the stannic oxide gas sensor may be listed as follows.

1. A long operating life and reasonable parameter stability
2. High reliability (in part because of the very simple associated electronics) leading to a low failure rate
3. Good resistance to corrosive gases
4. Robust construction and good mechanical strength
5. Any long-term drift is towards an increase in sensitivity (which contributes to "fail-safe" characteristics)
6. Inherently low cost (including the electronics), small and easy to handle
7. Little maintenance

Item 1 implies that stannic oxide, like most metal oxides, is chemically stable in normal (albeit polluted) atmospheres, which are oxidizing. It has been shown that there are various forms of drift with time, but given appropriate alarm limits, an instrument can work with little or no attention over at least a decade.

The high reliability and low failure rate of item 2 reflect the absence of moving parts, complex electronics, or possible liquid leakage (which can occur in electrochemical cells involving electrolytes).

Most corrosive gases found in the atmosphere, such as sulfur dioxide, do not affect the sensor at all, and item 3 implies that even for extreme concentrations, the sensor will recover in most cases.

Item 4 refers not only to the high mechanical strength inherent in the ceramic nature of the sensor material itself, but also to the modern robust housing and encapsulation containing it.

An important aspect of item 5 is that there have been few cases where the sensor has failed in such a manner that the failure remained undetected. The norm is for a marked increase in sensitivity to occur so that the situation becomes obvious. Clearly, failure of the associated electronics can take place, but well-known self-checking techniques exist which, though contributing to cost, are justified for commercial and industrial applications. For domestic applications, this is not normally necessary, and great reliability can be achieved via the circuit simplification which is made possible by the very high sensitivity of the sensor itself. This approach also makes for a very small and low-cost instrument, as noted under item 6.

As has been mentioned, an instrument based on a stannic oxide sensor needs little maintenance, so item 7 has been included to cover this. However, in extreme cases, such as where a monitor is used under critical commercial or industrial conditions, a program of inspection and recalibration should be carried out. This, however, is not by any means as onerous as in the case of a potentiostatic electrolysis gas sensor, for example, where periodical replacement of the electrolyte is needed, along with frequent checking for leaks and other problems associated with wet chemistry.

8.1.2. Disadvantages of the Stannic Oxide Gas Sensor

There are five prime features of the stannic oxide sensor which are regarded as disadvantageous. They are

1. Poor selectivity
2. Poor unit-to-unit consistency of performance (that is, large parameter tolerances)
3. Major dependencies on ambient temperature and humidity
4. Long stabilizing time after energization
5. Large power consumption (compared with electrochemical cells or ionization-type smoke detectors, for example)

The mechanisms which make possible the detection of gases by the stannic oxide sensor and which have been explained in the foregoing text, militate against selectivity, and item (1) reflects this fact. The natural processes are for all gases which can chemisorb onto the surface, or react with species already adsorbed, to do so. Hence, a problem exists which can, for example, produce false alarms resulting from build-up of alcohol, tobacco or paint fumes. As has been explained, some degree of selectivity can be conferred by the use of appropriate materials and additives,[1-3] and/or the careful choice of working temperatures or temperature cycling,[4,5] but this is never sufficiently marked for a single gas to be selected. However, a number of different (or multiple integrated) partially selective sensors may have their outputs combined in sophisticated electronic signal-processing networks to provide a useful analytical capability.[6-11] Item 2 is immediately relevant to the multisensor approach because the associated software is programmed to take account of the characteristics of each sensor, and any replacement will result in modification of those parameters, which will lead to the necessity for some reprogramming, or to "retraining" in the case of a neural network.

It is not only in sophisticated signal-processing apparatus that poor repeatability constitutes a problem, but even in simple gas alarms a sensitivity control is desirable for calibration, which should be carried out after aging. An alternative is for a manufacturer to perform calibration in order to offer sensors with individually defined characteristics and perhaps complete with customized load resistors.

There are, however, circumstances in which lack of selectivity is actually an asset. One example of this is the domestic gas alarm aimed at several gases including methane, propane, hydrogen, and carbon monoxide. An alarm which will cope with all these using only one sensor is perfectly feasible, and the lack of sensitivity in this case is meritorious!

The effects of ambient temperature and humidity have also been treated within the text, and the latter can be considered simply as an extraneous gas, which along with any other gases will affect sensitivity. If these environmental factors were to remain constant, however, repeatability to better than a percent of an alarm point could be expected. For example, if the lower explosive limit (LEL) concentration of methane is 50,000 ppm, and an alarm is set nominally at 5% of this (2500 ppm), then it should operate reliably somewhere between 4% (2000 ppm) and 6% (3000 ppm). Such conditions are rare in practical circumstances, but do exist, such as in operating theaters, where concentrations of anaesthetics such as halothane may be monitored.[12]

Ambient temperature variations result in changes in the physical condition of the gas to be detected, which is why thermistor compensation is never completely satisfactory.

Disadvantage 4 predominantly concerns the long-term transient described in Chapter 2: to fully stabilize a sensor requires energization for a week up to a month, depending on the particular sensor involved. After this period, a calibrated sensor can be expected to remain stable for continuous operation over more than a year. However, a long aging time is not strictly necessary for sensors destined for use in battery-operated portable apparatus used only over several hours for relatively crude measurements, including sensitive gas leak alarms.

Although potentially useful in fire detection[13] via carbon monoxide production, which occurs ahead of smoke production, current stannic oxide sensors cannot compete with ionization-type smoke detectors (which can operate for a year or more on one small battery), except where permanent mains operation is possible, and here, storage battery backup is often mandatory in case of mains failure. Item 5 addresses the comparatively high power required by the heater and is a major problem in the battery operation of contemporary sensors as considered further below.

8.2 POTENTIAL IMPROVEMENTS

Improvements in stannic oxide sensors and their associated electronics can be expected to follow three main paths. Firstly, advances in the sensor material and structure along with reduced power dissipation can be anticipated; secondly, more sophisticated methods of electronic drive and measurement will be employed; and thirdly, the concatenation of sensor and electronics in hybrid semiconductor assemblies will take place, leading to "smart sensors". Clearly, the latter expectation will result from success in the first two directions.

8.2.1 Improvements in the Sensor *per se*

Selectivity improvements based upon the inclusion of dopants, other than the noble metals as already described, are currently under investigation, as are more sophisticated filters having selective absorption and oxidizing capabilities. However, a further mechanism is available in the configuration of the active sensor material itself. For example, a gas which is easy to oxidize, such as carbon monoxide, will oxidize in the outer layers of a thick material and will not reach the inner parts at all. Consequently, the resulting resistance change is confined to these outer layers, and the bulk does not contribute, which limits the sensitivity. So, in principle, a thin film sensor would be appropriate under these circumstances, and, in fact, studies of multiple layer sensors are also relevant.[14] Combinations of these approaches should lead to considerably greater selectivity than has been hitherto achieved.

Problems involving poor repeatability, long stabilizing time, and drift towards greater sensitivity for certain applications have already been highlighted, and these are clearly connected with changes in catalytic activity which depend upon the fine structure of the metal oxide surface. As has also been mentioned, improvements may be attained by adding rhenium and vanadium, though the relevant mechanisms are far from clear. Hence, further basic studies of surface fine structure (including adsorbed species) are obviously needed before rational approaches to sensor improvement in this context can be formulated. It is to be expected that commercial requirements for improved sensors will eventually act as a driving force for comprehensive fundamental studies of this nature.[15]

Some potential sensor applications depend upon the development of low power versions, particularly in two areas. Firstly, the amount of power available via Zener barriers in hazardous areas is very limited, and for continuous operation in such areas, new low-power sensors would be advantageous. Currently, it is possible to fabricate ceramic sensors which require only about 70 mW, and these should prove satisfactory in such applications. The second relevant requirement is in battery-operated portable equipment, and here, the obvious prime factor is battery conservation.

Actually, 70 mW is about the lower limit of power consumption for the ceramic sensor, having regard to minimizing its thermal capacity, yet coping with heat loss via the supporting structures and leads,[16] and though thick film construction helps in this context, it is possible that thin film techniques may have greater advantages.

Convection losses and air flow also pose problems, and the latter can markedly affect the operating temperature of a small, low thermal capacity sensor. This highlights the problem of the increasing effects of ambient temperature and air-flow on the operating temperature as sensors become smaller and indicates that an electronic driver circuit would benefit from feedback via a temperature sensor (such as a thermistor) buried in the stannic oxide ceramic.

8.2.2 Improvements in Sensor-Circuit Combinations

The combination of sensor technology and electronics can lead to some very promising developments, and a good example of this is the use of partially selective sensors with signal-processing electronics to result in the analytical instrument, or "electronic nose" previously mentioned. Another example is afforded by the use of rate-of-change control. Here, if a sensor is drifting slowly due to ambient temperature and RH changes in an atmosphere, the value of its most recent resistance can be stored and called R_0. Then, if the rate-of-change of resistance rises or falls rapidly, signaling a major change in gas concentration, a new resistance is stored as R_S after settling. Such a method can be used to combat apparently poor stability and repeatability, and has been used to control an air cleaning system[17] by switching on equipment when a defined minimum difference between R_0 and R_S has been recorded. Clearly, appropriate associated electronics could also be used to calculate and display the resistance ratio for the sensor R_S/R_0 while minimizing drift errors.

The maintenance of a constant operating temperature has so far been assumed to be performed via the positive temperature coefficient characteristic of a metallic heater working at a constant applied voltage. This actually constitutes a case of negative feedback wherein the temperature-dependent resistance of the heater itself is used to control its own power dissipation and hence the temperature reached. A typical circuit using this approach is depicted in Figure 6.4.

Better temperature control would be achieved if amplification were inserted into the feedback loop and a simple way of doing this is illustrated in Figure 8.1. Here, the heater resistance is compared with a reference resistance R_R within a Wheatstone bridge configuration. The offset or balance voltage of the bridge is applied to the inputs of a difference amplifier which drives a controlled current source.[18] Given connections of the correct polarity, this source is continuously adjusted so as to maintain this balance voltage at almost zero, which implies that the heater resistance, and hence the operating temperature, remains almost constant. The desired heater resistance and hence operating temperature may be preset by an appropriate potentiometer adjustment as is also shown in Figure 8.1. This form of heater drive, involving a small-power field-effect transistor, is employed by Capteur Ltd. for its range of second-generation sensors.[19]

As an alternative to using the resistance of the heater itself as a temperature sensor, an independent transducer might be buried in the active material to provide a signal for the feedback loop. Unfortunately, thermistors, which are very sensitive, are not in general capable of working at the operating temperatures of most stannic oxide gas sensors. Hence, platinum wire resistance thermometers or thermocouples might be mandatory, which implies that the use of the heater resistance itself will hardly be improved upon.

Again referring to Figure 8.1, it will be appreciated that the value of R_R cannot be significantly less than the value of the heater resistance itself, for it

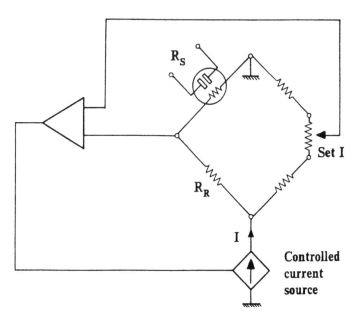

FIGURE 1.   A method of controlling sensor heater temperature.

must carry the entire heater current: hence, it will consume comparable power. This will usually be unimportant in mains-operated instruments, but for battery operation, it may be untenable. This is also the case for circuits such as that of Figure 6.4, where equally significant power can be consumed in the voltage controller. One answer to this dilemma is to replace continuous heater current with pulsed current such as from a pulse-width-modulated power supply. These are already well-known for controlling average heater currents in many applications, and such techniques have recently been applied to gas sensors working down to 14 mW.[20] Here, however, previously noted reservations regarding heat loss by conduction, and natural and forced convection, are still important.

A sensor capable of working at room temperature would be ideal, but though efforts are being made to realize such a device, there are formidable problems. For example, various atmospheric pollutants, including the species to be detected, will adsorb cumulatively, whereas normally a high-temperature sensor is either continuously or intermittently purged. At best, a room-temperature sensor based on present technology would be very slow, and at worst, rendered inoperative with the passage of time. None of this, however, precludes technologies not yet publicly reported, and it is certain that a major market exists in many areas, including fire detection via the early release of carbon monoxide[21] and hydrogen,[22] but without the necessity for the provision of comparatively high heater power.

8.2.3 'Intelligent' Sensors

An "intelligent" sensor refers to the concatenation of the sensor itself along with its driving and measuring circuits in a single package to form a hybrid integrated circuit. The electronics does not constitute a problem because application-specific integrated circuits (ASICs) are well known, and can range from highly expensive full custom integrated circuits to very much cheaper semi-custom devices. The latter usually require the deposition of only a layer of metalization and are available in the form of nondedicated structures ranging from single transistors to both logic cells and linear "tiles". Hence, pulse-width-modulation drivers and operational amplifier based measuring circuits present no major difficulties.

The main problem is that of heat generation by the sensor, which could easily heat up the electronics to deleterious levels.[23] Hence, either very low power sensors — perhaps in thin film form[24] — are needed, and/or appropriate heat dissipation techniques must be applied. The former point has been addressed above, as has the question of sensor stability, which is also of concern.

Clearly, there are many areas of investigation remaining to be covered in stannic oxide gas sensor technology, and having regard to the current frequency of publication, progress will be made in all of them, plus some so far not envisaged.

REFERENCES

1. Coles, G.S.V., Williams, G., and Smith, B., Selectivity studies on tin oxide-based semiconductor gas sensors, *Sensors Actuators,* B3, 7, 1991.
2. Yamazoe, N., New approaches for improving semiconductor gas sensors, *Sensors Actuators,* 5B, 7, 1991.
3. Moseley, P.T., Materials selection for semiconductor gas sensors, *Sensors Actuators,* 6B , 149, 1992.
4. Sears, W.M., Colbow, K., Slamka, R., and Consador, F., Surface adsorbtion and gas consumption in restricted flow, thermally driven, gas sensors, *Semicond. Sci. Technol.,* 5, 45, 1990.
5. Hirakana, Y., Abe, and T., and Murata, H., Gas-dependent response in the temperature transient of SnO_2 gas sensors, *Sensors Actuators,* 9B, 177, 1992.
6. Ikegami, A. et al., Olfactory detection using integrated sensor, Proc. Transducers '85, 3rd Int. Conf. Solid-state Sensors Actuators, Philadelphia, 1985, 136.
7. Shurmer, H.V., Gardener, J.W., and Chan, H.T., The application of discrimination techniques to alcohol and tobacco using tin oxide sensors, *Sensors Actuators,* 18, 361, 1989.
8. Shurmer, H.V., An electronic nose: a sensitive and discriminating substitute for a mammalian olfactory system, Proc. IEE Part G, Ccts. Syst. 1990, 138.
9. Gardner, J.W., Hines, E.L., and Tang, H.C., Detection of vapours and odours from a multisensor array using pattern-recognition techniques, *Sensors Actuators,* Part 1: 4B, 109, 1991; Part 2, 9B, 9, 1992.
10. Nakamoto, T., Tagaki, H., Utsumi, S., and Moriizumi, T., Gas/odor identification by semiconductor gas-sensor array and an artificial neural-network circuit, *Sensors Actuators,* 8B, 181, 1992.

11. Carey, W.P. and Yee, S.S., Calibration of non-linear solid-state sensor arrays using multi-variate regression techniques, *Sensors Actuators,* 9B, 113, 1992.
12. Lee, A. and Snowdon, S.L., Semiconductors in anaesthetic vapour analysis, *Anaesthesia,* 33, 352, 1978.
13. Wagner, J.P., Fookson, A., and May, N., Performance characteristics of semiconductor gas sensors under pyrolytic, flaming and smouldering combustion conditions, *J. Fire Flammability,* 7, 71, 1976.
14. Shimizu, Y. et al., Gas-sensing properties of multilayered thick film SnO_2 gas sensors, Proc. 3rd. Int. Meet. Chem. Sensors, Cleveland, 1990, 120.
15. Chiorino, A., Bocuzzi, F., and Ghiotti, G., Surface chemistry and electronic effects of O_2, NO and NO/O_2 on SnO_2, *Sensors Actuators,* 5B, 189, 1991.
16. Figaro Engineering Inc. data sheets on the TGS 501 sensor.
17. Matsumoto, T. et al., Air quality sensor, Dig. 7th Chem. Sens. Symp., Jpn. Assoc. Chem. Sens., Electrochem. Soc. Jpn., Saitama, 1988, 169.
18. Bennamar, N. and Maskell, W.C., Temperature control of thick film heaters, *J. Phys. E. Sci. Instrum.,* 22, 933, 1989.
19. Capteur Sensors & Analysers Ltd., Product Guide, May 1993.
20. Amamoto, T., et al., Pulse drive semiconductor gas sensor, Digest of 13th Chem. Sens. Symp., Chem sens. (J. Jpn. Assoc. Chem. Sens., Electrochem. Soc. Jpn.), 7, suppl. B, 1991, 117.
21. Oyabu, T., A simple type of fire and gas leak prevention system using tin oxide gas sensors, *Sensors & Actuators,* 5B, 227, 1992.
22. Amamoto, T. et al., A fire detection experiment in a wooden house by SnO_2 semiconductor gas sensor, *Sensors & Actuators,* B1, 226, 1990.
23. Manaka, J. et al., Low electric power and high speed response micro gas sensor, *Proc. 32nd Meet., Jpn. Soc. Appl. Phys.,* 765, 1985.
24. Schierbaum, K. D., Weimar, U., and Gopel, W., Comparison of ceramic, thick-film and thin-film chemical sensors based upon SnO_2, *Sensors & Actuators,* 7B, 709, 1992.

APPENDIX

Testing and Characterization

A.1   ENVIRONMENTAL SENSITIVITY OF SENSORS

In the foregoing text, the responses of stannic oxide sensors to various stimuli have been described, and these may now be listed as a prelude to considering methods of testing and characterization. They are

1.   Gas concentration (of the gas it is desired to detect)
2.   Other gases
3.   Ambient humidity
4.   Ambient temperature
5.   Air velocity
6.   Sensor element temperature
7.   Time

Of these, only item 1 implies a wanted response, the others being undesirable; indeed in the absence of these others, the stannic oxide sensor would be an excellent concentration-monitoring transducer down to very low parts per million levels.

Throughout the text, graphs showing the responses of various forms of sensor to a selection of gases have been given, often as functions of the temperature of the sensor element. In this context, items 2 and 3 are similar because water vapor can be regarded as simply another gas.

Knowing that the rates of chemical reactions increase with temperature, it is not surprising that ambient temperature, as well as the heater-defined temperature of the active material, affects sensitivity. In fact, this is exacerbated by the influence of the ambient temperature on the rate of cooling of the sensor, and this tends to be more marked in the cases of those sensors (such as cool-running carbon monoxide-sensitive types or those miniature versions for battery operation) which receive comparatively low power levels from their heaters.

This effect can be minimized by incorporating a temperature sensor (which may be the heater itself) and using it within a feedback loop to control the heater power, as has been seen.

The final environmental factor is air velocity over the sensor encapsulation, for this provides a cooling effect which reduces the operating temperature, and so is also particularly important for low-temperature sensors. However, it can be minimized by the use of a properly designed housing.

These are the factors which must be taken into account when characterizing a sensor, so that any apparatus intended for this purpose must be capable of generating an environment which can be controlled in terms of all of them concurrently.[1-3]

Item 6 refers to the target temperature of the bead, and this is a function of the balance of power applied against the rate of heat loss via conduction and convection. In the absence of draughts and ambient temperature changes (and to a lesser extent, exothermic reactions on the active material surface), a constant heater power would lead to a constant sensor element temperature. Usually, however, a constant voltage supply is used which is satisfactory for most purposes, especially when it is recalled that a metal heater has a positive temperature coefficient, so that there is some degree of inherent temperature control.

Finally, as has been shown, stannic oxide sensors have both short- and long-time dependencies. This means that the sensor(s) under test must be allowed to stabilize before characterization, and though this applies initially to the short-term switch-on transient, the long-term or aging transient should be allowed to elapse (or the sensors must be pre-aged) for sensors destined for use in the more critical situations.

A.2   THE REALIZATION OF LOW GAS CONCENTRATIONS

Commercial gas blenders are well-known and will provide percentage mixtures of various gases accurately and repeatably. Such mixtures are usually adequate for the checking and testing of detectors for lower explosive limit (LEL) levels of combustible gases in air, for example, but not for the much lower tolerance limit value (TLV) concentrations of toxic gases. They can, however, be used to further dilute premixed gases to provide such low-level concentrations, but the results depend on the accuracy of the pre-mix and whether it degrades over time. Carbon monoxide with a TLV of 50 ppm (or 35 ppm in the U.S.) is a case in point, and some other gases have TLVs even below 1 ppm.

At these extremely low levels, it is possible to use permeation tubes which are made from PTFE and contain liquid gas. This gas permeates through the walls of the tubes at rates defined by the ambient temperature, so that if placed in a constant temperature airstream, a known gas concentration can be produced which is a function of the velocity of that airstream.

Unfortunately, the concentrations available by this method are inevitably very low indeed, and solid state gas sensors are not very useful in such regions at present because the effects of the environmental factors quoted above tend to be greater than those of the detected gas.

Higher concentrations, from perhaps 5 to 5000 ppm, can be conveniently obtained by using a gas syringe. Here, a chamber is used into which clean air can be introduced and which has a rubber septum incorporated. A pure gas is drawn into a calibrated gas syringe, after which the needle is pushed through the septum and the pure gas is injected into the known — and much larger — volume of clean air. This is then mixed using an internal fan, and any sensors within the chamber will react appropriately and their conductances may then be measured. (This procedure must be performed with dispatch if the chamber is small; otherwise the gas will be consumed at the catalytic surfaces of the sensors and the concentration will fall.)

This is a "batch" process and is obviously useful for determining the sensitivities of sensors (R_a/R_s) since their resistances in both clean and contaminated air can be easily measured. However, such measurements are inevitably taken under uncontrolled conditions of ambient temperature and humidity, but this may be acceptable if the sensors are intended for use in simple gas alarms, such as domestic devices, for these can be set at sufficiently low alarm levels that they will operate safely even if the environmental conditions are such as to reduce the apparent sensitivity. However, industrial sensors for use in monitors must be characterized more comprehensively, as detailed below.

A.2.1 Gas Sensitivity Characterization

Precisely how a sensor characteristic is to be displayed depends upon the use to which that characteristic is to be put. However, the obvious parameter to measure is the conductance because this is both easy to determine and is the most useful parameter, as noted in Section 1.5. Furthermore, such measurements are easily manipulated to provide logarithmic resistance ratios and sensitivity plots where necessary.[4]

In principle, each sensor will exhibit a "sensitivity surface" which is a three-dimensional plot of the sensor conductance in terms of gas concentration, ambient temperature, and humidity measured under normal operating conditions. In practice, though all these parameters can be concurrently controlled (as will be seen below), this is a rather extreme approach. It is more practical to determine the eventual application of a sensor and characterize it accordingly, so minimizing the cost.

A.3 THE PROBLEMS OF AMBIENT TEMPERATURE AND HUMIDITY

The maximum ranges of temperature and RH over which a sensor needs to be characterized are usually from 0 to 50°C and 5 to 95% RH. Neither range is difficult to attain separately, but the concurrent extremes present considerable practical difficulties. For example, at high humidities and low temperatures, water vapor easily condenses onto the surfaces of the test chamber and its internal components. Also, because RH is itself a function of temperature, the necessary humidifying apparatus must be able to cope with the resultant increased need for water vapor at high temperatures.

The test chamber temperature may be controlled via both an internal heater and cooler. The former may be simply a resistive element while the latter may be a small refrigerator cooling coil. An internal calibrated temperature sensor may be used to control the operation of both via an appropriate feedback network. At temperatures well above or well below ambient, the heater or cooler will operate alone, and good control is easy to achieve; however near ambient, where thermal overshoot becomes important, adequate control can become problematical.

Humidification can be achieved by bubbling air through water (preferably warm), and it will be realized that this process can itself affect the temperature of the chamber. Dehumidification can, in principle, be achieved via a condensation tube, possibly operated via the existing refrigeration system, but this again can affect the chamber temperature. A better solution is to introduce dry air from a cylinder and humidify it as required.

A.3.1 An Environmental Test Chamber

Figure 1 shows an appropriate test chamber in diagrammatic form.[3] This consists of two chambers, one inside the other with a copper separator, a form of construction which minimizes the condensation problem via rapid temperature equalization. The outer (lower) chamber contains the heater and cooler, while the inner (upper) chamber contains a "bubbler" bottle of distilled water. Figure 1 shows this arrangement, but omits the bubbler for reasons of clarity. It does, however, show an electric fan mounted in the outer chamber which is magnetically coupled to a slave fan in the inner chamber, so avoiding any temperature effects in this inner chamber due to fan coil heating. In operation, the temperature of the inner chamber is defined by that of the copper separator, while the RH is defined internally. It has been shown that such an arrangement can be operated using a small computer, so that runs of temperature at a fixed value of RH — and *vice versa* — can be carried out automatically.

The computer can also control the gas injection sequence as depicted in Figure 2. Here, dry air flows into the inner (test) chamber while a defined gas flow is continuously vented. The computer first isolates the air flow to the test chamber, then operates a two-way valve to allow the gas flow to enter the test chamber for a specific time. This defined volume of gas is mixed with the air by the internal fan while the temperature and RH are held constant. By purging the test chamber and repeating the procedure for a series of different environmental conditions, a set of sensor conductance measurements can be automatically carried out and plotted in the most appropriate form.[5]

This latter gas injection method can obviously also be used with a simple "ambient test chamber" for less stringent testing purposes.

A.4   USER CALIBRATION

Although complete preset and tested gas alarms are available,[6] it is advisable for the individual user to be able to check them at intervals having regard to the long-term drift problems detailed earlier and, of course, the effects of any dirt or even damage. This may be done, albeit crudely, using aerosol cans containing a calibration gas mix which may be applied to the instrument via a simple tube, or preferably a lightweight plastic hood. Such tests are adequate for the checking of explosive gas alarms which usually operate near 5% of LEL, but are much less satisfactory for toxic gas alarms which typically work in the tens of the parts per million range.

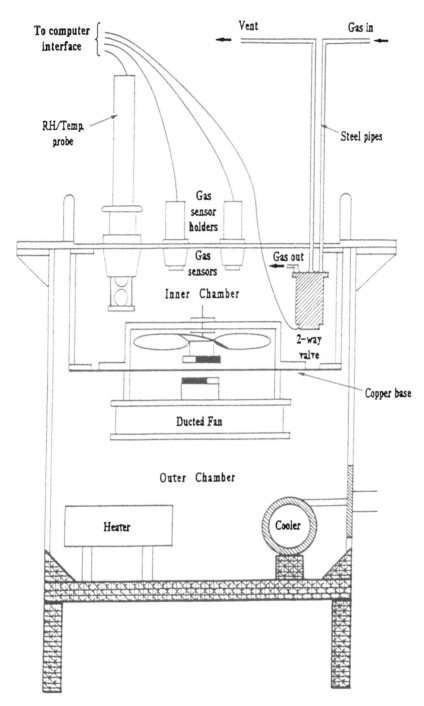

FIGURE 1. Environmental test chamber arrangement (humidifier in inner chamber omitted for clarity).

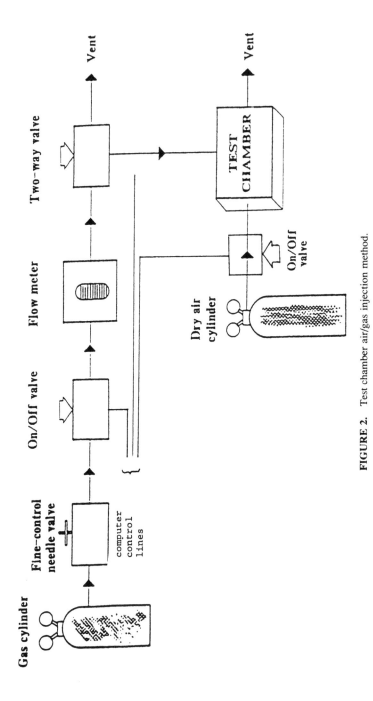

FIGURE 2. Test chamber air/gas injection method.

In the case of major users such as large corporate bodies, city authorities, insurance or voluntary approval organizations, more sophisticated tests may be performed on sample instruments. For example, the Japan Gas Appliance Inspection Association recommends an "accelerated" test procedure in which an alarm instrument is maintained at 35°C and 60% RH for 10 days. During this period, hydrogen is introduced to give a concentration of 500 ppm for half-an-hour on a twice-daily basis. After the 10 days, the alarm level is checked and instruments which alarm at or below 400 ppm of hydrogen or 500 ppm of ethanol are rejected. The theory here is that this procedure identifies instruments which would fail had they been continuously activated under more normal conditions for a 5-year period.

This is not only an example of an accelerated test procedure, but also implies that a plethora of test procedures could appear around the world unless major bodies representing large populations (such as the European Community) produce mandatory requirements, which would be a more satisfactory situation.

REFERENCES

1. Bott, B. and Jones, A., A microcomputer controlled system for characterising semiconductor gas sensors, *Lab. Pract.*, 32, 6, 80, 1983.
2. Demarne, V., Grisel, A., Sanjines, K., and Levy, F., Integrated semiconductor gas sensors evaluation with an automatic test system, *Sensors Actuators*, B1, 87, 1990.
3. Harvey, I., Coles, G.S.V., and Watson, J., The development of an automatic test chamber for the characterization of gas sensors, *Sensors Actuators*, 16, 393, 1989.
4. Watson, J., A note on the electrical characterization of solid-state gas sensors, *Sensors Actuators*, B8, 173, 1992.
5. Mousa-Bahia, A.A., Coles, G.S.V., and Watson, J., A gas injector for an automatic environmental test chamber for the characterisation of gas sensors, *Sensors Actuators*, B12, 142, 1993.
6. Figaro Engineering Inc., Manufacturers guide to the production of TGS gas detectors including illustrated examples of recommended calibration procedures.

INDEX

9 780367 449513